Finishes

MITCHELL'S BUILDING SERIES

Finishes

Fourth edition

YVONNE DEAN

LONGMAN

Addison Wesley Longman Limited
Edinburgh Gate, Burnt Mill, Harlow
Essex CM20 2JE, England
and Associated Companies throughout the world

First published by The Mitchell Publishing Company Limited 1971
Second edition 1979
Third edition 1989
Fourth edition published by Addison Wesley Longman Limited 1996

British Library Cataloguing in Publication Data
A catalogue entry for this title is available from the British Library

ISBN 0-582-25877-4

Set by 4 in Times 9/11 and Melior
Produced by Longman Singapore Publishers (Pte) Ltd
Printed in Singapore

Contents

Acknowledgements

I would like to thank the following for their help in the preparation of this book:

Special thanks to Dr Ian Boustead of the Open University and Dr Barry Pepper of Tottenham College of Technology for reading the book and making valuable comments on the contents, Rosemary Handyside who acted for a short time as my Research Assistant on some developmental areas of the book and Teresa Hoskyns who helped check all standards and gathered information for the fourth edition.

Kim Davies and Mark Edwards, and the Media Services Centre of the Polytechnic of North London for their help in the preparation of photographic material (Figures 2.14–17, 3.4, 3.11, 4.3, 4.5, 5.5, 5.8, 5.11, 5.21–23, 5.25 and 6.1); Michael Marlowe for his photographs (Figures 3.7, 5.14 and 5.18); Steve Keates for his photographs (Figures 1.1 and 8.1) and his help in the conversion of colour slide material to black and white photographs; and Julia Dwyer for her support and her help in the preparation of illustrative material (Figures 2.2–7, 2.9, 2.10, 3.1–3, 3.5, 4.1, 4.2, 5.20, 7.1, 7.3 and 7.5).

In memory of Derek Osbourn RIBA, formerly Head of the Department of Environmental Design at the Polytechnic of North London, for his encouargement.

Thelma Nye, formerly of Batsford, for her editorial guidance; Alan Everett for his help and advice; Peter Bowbeer of the University of North London library; and Barbour Index for the loan of their microfiche. Lastly, I still acknowledge the interest and reactions given by students to my lectures, which influence the way text and visual material is organised.

Yvonne Dean, London 1996

The publishers are grateful to Leonard Hill, Blackie and Son for permission to reproduce copyright material from I.J. Mcholm 1983 *Ceramic Science for Materials Technologists*.

Extracts from British Standards are reproduced with the permission of BSI. Complete copies can be obtained by post from BSI Customer Services, 389 Chiswick High Road, London W4 4AL; Telephone: 0181 996 7000.

The American Society for Testing and Materials (ASTM) is referenced in this book. Further information can be obtained from: 1916 Race Street, Philadelphia, Pennsylvania 19103, USA.

1 Introduction

Although this volume is concerned with 'Finishes' as a subject, a title such as 'The Visible Appearance of the Building Fabric', or even 'The External Skin and its Behavioural Properties and Characteristics' might be more appropriate. The use of finishes is often regarded as a separate and final application to the fabric of the building, sometimes even the last part of the building to be specified. This may mean finishes are subject to compromise in their quality by late cost-control exercises. However, finishes represent the true boundary of a building, the face for immediate contact and, more critically, should be seen as the first line of defence for protection of the fundamental fabric of the building. Some finishing elements are a good deal more substantial than others and are often part of the solid external skin, and distinct from further applied coverings or coatings. Consequently finishes are one area of the building industry that is continually involved in replacing materials in a way that could be seen as sacrificial.

Figure 1.1 Peuterey Ridge and Aiguille Noir, Mont Blanc, September 1981. The scale of exposure that our buildings have to cope with is easily forgotten, and our climatic extremes are often peaks in an undulating chart of annual variations. If we catered for the worst possible conditions, our buildings would be better equipped for survival. (*Photograph*: Steve Keates.)

It is still appropriate and convenient to think of thick and thin finishes or, to be more exact, macro and micro finishes. Micro finishes refer to paints, films, and other surface finishes whose substance and thickness is difficult or impossible to determine or measure accurately with the naked eye. The basic unit used is 10^{-6} m and is referred to as a 'micron'. Macro finishes are large-scale materials that can be seen with the naked eye. A reasonable estimate can be given of their chief characteristics and dimensions, which will be in millimetres and metres. Macro finishes still have a detailed microstructure that determines their properties and behaviour.

Both categories of finish are used externally and internally and there are different environments for both categories. In fact, the range of temperatures and conditions in which we expect materials not only to survive, but also to remain stable, is extraordinary. Even in the British Isles, human beings cannot survive for long at the extremes of these basic parameters if totally exposed.

We also rely to a great extent on our own observational experience to judge conditions and to make an appropriate choice of materials, but our individual views as designers, architects, contractors and practitioners are limited.

The internal movements of molten magma in the upper mantle, together with the shifting of tectonic plates, contribute to distort the thin outer skin of the Earth's continental and oceanic crust into mountains, volcanoes and other features. The Earth, under the continual action of rain and wind-borne dust, would become smooth and polished like a well worn pebble on a beach, were it not for this internal disruption that continually changes the profile of its surface.

The fabric of our buildings is also exposed continually to a cyclical action of wind, rain, cold and heat which in the long term erodes naturally all physical features on Earth. Chemistry too can erode an outer surface by encouraging the deterioration of materials.

Deterioration, however, is an inaccurate description of this process: materials alter to become chemically stable in our atmosphere. This is usually a state described in thermodynamics as one of 'low free energy'. Complex man-made products usually have sophisticated structures, and their very existence depends on a too-easily altered energy balance.

In this century many finishes have been affected by higher levels of pollution and there has been great concern with the effect of acid rain. Emissions of acid into the atmosphere seemed to peak in 1970 and there has now been a reduction of almost 30% in the acidity of some lakes. De-sulphurisation of waste gases from power stations has reduced emissions of sulphur dioxide and the British Govt did agree with the EC to cuts totalling some 60% by 1993. The fall out of nitric acid is now more significant as a contributor to acid rain than sulphur dioxide. The most evident indicator of this type of pollution is 'sick tree syndrome' where the needles of conifers change colour and die. However, this type of tree sickness may also be attributable to the effects of magnesium deficiency from the long hot dry summers of the early 80s. Lack of rain effected the release of this mineral from the forest floor into the earth and it stunted root growth. The accumulation of acids, sulphate and nitrates in the soil since the industrial revolution has certainly also affected root growth.[1]

As building professionals, we use a technology that is generally inappropriate to deal with the very real and large-scale events that are commonplace in our environment and we construct without consideration for the effects of weather, climatic change, sunshine or temperature. Physics, earth sciences and chemistry have become so remote and so specialized that the experience gained in these fields is usually not easily transferable and therefore not often used in the low technology field of building. We also use high technology solutions to help us out of awkward situations where our building technology is inadequate. In fact we are really seeking solutions that are sophisticated in terms of their complexity, and yet are an abuse of some fundamental principles in science. This can hasten decay in buildings and also use up primary resources, not only in terms of the raw materials extracted and refined for construction and repair, but also in terms of energy used in their processing and treatment to keep them in a stable state.

Environmental legislation

In the last decade our general concern with the effects of manufacturing industries and consequential environmental impact has been followed by global, European and national agreements.[2] The first United Nations Conferences on the Environment and Development resulted in the Rio Declaration to assist environmental action and future development. Agenda 21 set out a plan of action for the 21st century which emphasizes the need for energy-efficient technologies, renewable energy sources and the reduction of pollution levels.[3] Policy is fast being supplemented by action that satisfies public concern. Legislation is becoming increasingly difficult to keep up with and many new directives are now being issued through the EC. Individuals want to show they are complying with new directives even though enforcement is minimal. Dissemination of new material through the technical press is slow and the best sources for environmental information are a network of European Documentation Centres in Britain. These include the main Science and Technical reference library of the British Library in Holborn, selected Universities and major Chambers of Commerce. There is the *Manual of Environmental Policy: the EC and Britain* which is updated twice a year. This includes all the latest EC environmental directives with commentaries on their development and effect.

1.1 The importance of specification

The correct specification of finishes is vital to the durability and appropriate use of a given material.

Most finishes applied to the surface of buildings fall into two major categories: polymers and ceramics. If the nature of these materials is understood then their correct application and use will be easier to specify. Consequently this book groups finishes into categories with similar characteristics as normally defined in materials science. It is unreasonable to suppose that designers can memorize the properties of different products individually. A general understanding of polymeric materials and how they behave will assist in the specification and use of finishes such as paint, flooring compounds, roofing compounds and adhesives. Similarly, an understanding of the properties of ceramic materials and how they fail helps in determining their use in the design of fixings and in carefully supervising during installation.

In the past textbooks on this subject tended to be reference books. They dealt with the history, practice and usage of materials and their finishes. These books, usually a three or four volume set, encompassed the state of the art in building. They were also prescriptive, i.e. they were able to detail the exact materials in use at the time with their limited and proven application.

In the latter part of the twentieth century we have seen a great increase in the materials and products available. Descriptions of all these products can be obtained from publications produced and updated annually by the Architectural Press, Barbour Microfile, RIBA Publications Ltd and, more recently, the Building Technical File. These are core sources and should always be used as the starting point for investigation. The Building Centre can be used for queries on products and the Building Research Establishment for advice and diagnosis on building failures. The Design Centre offers advice on innovation in materials technology at its new Materials Centre. To ensure that there are no omissions in a specification, the National Building Specification now requires a standard list of documentation.

A textbook today can provide a broad background of the scientific principles. Armed with an understanding of materials and the mechanics of their deterioration, a diagnostic attitude to the use of materials can be developed by the student. This should lead to a greater sensitivity in the specification of finishes and their uses.

1.2 A strategy for specification

At some point every designer, architect or contractor deals with the specification of materials. Architects and designers have very specific responsibilities under the terms of their contractual arrangements with clients. These are further reinforced in standard building contracts. This responsibility cannot be delegated without agreement and designers must remain in control of the situation. It is the contractors' or subcontractors' responsibility to carry out the works as specified. Clauses of contracts in current use should be very carefully checked.

Far too often a document is written using information compiled for a different project. This information is repeated without checking it at source. Or information such as numbers for British Standards is copied without confirming its relevant or validity, or even whether a BS number is still available and not superseded. Instead of attempting to find a master document to imitate, it is far more useful to develop a clear strategy about specifying materials and to apply a simple set of parameters to each material or component in order to build up a reservoir of information. Existing specifications can then be used as checklists but care should always be taken. No specification will be sufficiently exhaustive to deal with every aspect of a particular building. The quotation of a British Standard may on the other hand be too broad for a specific requirement. Even the National Building Specification may still need clarification of particular items. If the following three steps are taken for every component in a building, a reasonable specification can be compiled:

1. The nature and composition of a material or component should be specified by reference to available and current British or other recognized standards and/or Agrément Certificates, and in conjunction with manufacturers' literature if applicable.
2. The method of fixing or placing these materials in

position should be specified and the use of codes of practice and their relevant parts clearly outlined.

3. The method of protection or finish to the materials should be stated and specified together with their fixings. This may reveal that fixings should not have an applied finish but the nature of their fixings needs more careful specification.

The specifier must be organized: as the design or decision is made, information is kept perhaps in a loose-leaf binder or on a disc on a computer. The information should be concise and simply stated. There is no need to be legalistic, just be precise and clear about everything to be used. If this is done systematically you will have a good specification document and the source of information for Bills of Quantities. Errors and consequent failures in building are often associated with poor specification. Often the process of specifying materials is the last activity in the design process. In order to expedite documentation at the end of a project, fast decision-making may ultimately commit the designer and the contractor to unsound detailing and predictable repercussions. A good specification is inseparable from the intentions of the original designer backed up by technologically sound information.

1.3 How to use this book

This book is designed to be more than a reference volume. When considering a particular finishing material, it should first of all be identified as either a polymeric, ceramic, metallic or composite material. The general introduction in the relevant section provides the basic principles about the behaviour of the material. Reference can then be made to applications of individual materials. There are overlaps between these areas.

When using this book as a basic text all the introductory sections should be read first to give a broad base for understanding the subject. The glossary sections are expansive explanations of terms used in the relevant sections and should be treated as an informative dictionary.

For those with particular needs on the precise specification of materials, direction is given to source material such as British Standards and other specification documents.

Standards are often quoted by number without being scrutinized first by the specifier. If they were, practitioners would possibly find that the very clear directives and advice given in these documents conflicts with their own clauses and choice of specification.

Often the names and numbers of particular standards are quoted in their entirety by practioners when only a small section is relevant.

Up to date British Standards and Codes of Practice should be available for constant reference. It is confusing to quote to contractors entire standards when it is unnecessary and it is more likely that the guidance given will be ignored. The misuse of reference material also implies a lack of knowledge of the terms of the standard on the part of the specifier.

Moreover the reference to a British Standard does not absolve the specifier from legal responsibilities. A phrase often quoted in British Standards is:

'Compliance with a British Standard does not of itself confer immunity from legal obligations.'

Photographs

All the photographs have been selected to illustrate particular points. Some photos have a 'dot' which measures 8 mm in diameter and conveys an idea of scale.

Notes

1 New Scientist 15th September 1990.
2 *The Earth Summit Agreements: A Guide and Assessment* Earthscan Publications Ltd, London, 1993.
3 Chapter 7 of Agenda 21 *Promoting Sustainable Human Settlement Development*.

2 Polymeric materials

2.1 Introduction

Many polymers used in the building trade are commonly referred to as *plastics*. Plastic is really a descriptive term. A polymer (many monomers or 'mers') is a chain of many hundreds of thousands of individual molecules (Fig. 2.2). A single molecule of a polymer has an enormous molecular mass. If some basic knowledge can be grasped of the characteristics of these long-chain molecules, it can help us to understand how paints and other surface films behave, or at the very least give us a healthy respect for their properties and usage.

Polymerization is the process by which the individual monomers react chemically to form a polymer which exists in a state of lower free energy than the individual monomers. Sometimes compounds combine freely when the right conditions prevail, to create the thermodynamically favoured polymer. Often there has to be an input of energy: this can be in the form of directly applied heat, ultraviolet light, exothermic reactions from other local chemical reactions or (more unusually) from high-energy bombardment by electrons or gamma rays. Sometimes we can see this polymerization process happening in ordinary domestic situations. For example, perfume left in a bottle can, under the action of ultra-violet light, change from its clear liquid form to a sticky mass, from individual monomers of *terpene* into a resinous mass of polymers known as *polyterpene*.

Example Ethylene is the basic building block of the polymer polethylene, or *polythene* for short. In the middle of the ethylene molecule is a carbon double bond which can be broken if sufficient energy is supplied. Although individual ethylene molecules can react together in the right conditions to form

polythene, the process of polymerization is much faster if a catalyst such as chromium oxide or an unstable organic peroxide is added (Fig. 2.3). The

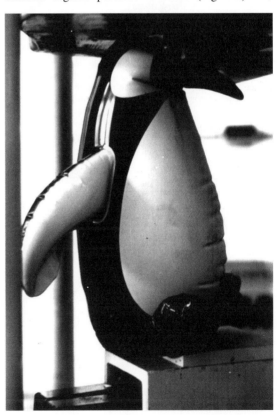

Figure 2.1 Inflatable penguin in an Avery-Dennison testing machine. Given this stressful situation most designers would accurately predict the outcome of loading to the limit. The stresses imposed on ordinary paint films and polymeric materials may not be visually obvious, but with some additional basic knowledge designers can be aware of, and cater for, equally stressful conditions.

chemically unstable peroxide molecule decomposes readily to form a highly reactive species called a *radical*. This attacks the double bond of ethylene, forming a new bond to one carbon of the ethylene molecule, but leaving the other carbon with an unpaired electron which can then attack another molecule of ethylene to form another new bond, and so on. The chain of ethylene molecules increases in length to many hundreds or thousands of sub-units. Finally, the chain reaction stops when the catalyst molecule is regenerated by a different reaction at the starting end of the polymer molecule.

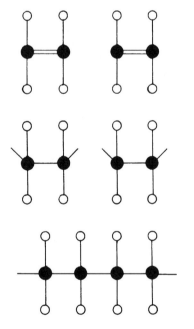

Figure 2.2   A polymer chain. The circles represent repeating monomer units (molecules), not individual atoms.

Figure 2.3   Ethylene. An ethylene molecule with the carbon double bonds broken by an input of energy, then spontaneously linking with another altered ethylene monomer to form a polymer chain.

The length of the polymer chain varies according to how the chain terminates. There are several ways in which this can be achieved. The manipulation of the chemistry at this level is known as *polymer architecture* as the planning and control of these large chemical compounds has to be so exact. The length of chain is important: it alters mechanical and chemical properties, affecting how the material deteriorates over time. This is particularly important in the processing of synthetic rubbers, where excessive chain length can make the material highly viscous, over-stiff and difficult to process. Long molecular chains can give the finished product a roughness which may not be desirable.

There are three fundamental categories of polymers: *thermoplastics, thermosets* and *elastomers* (Fig. 2.4). It is important to grasp the chief characteristics of these polymers as often finishes combine polymers with a variety of characteristics giving very different properties and hence applications.

Thermoplastics

Thermoplastics are long-chain polymers which are linear (i.e. without side branches) but have some degree of chemical bonding to adjacent chains (Fig. 2.5). It may be weak van der Waals' bonding (attractive forces between molecules dependent on electronic configuration) or hydrogen bonding (electrostatic attraction between a hydrogen atom and another element). Both forms of bonding can be weakened by heating or by induced stress. Thermoplastics are the most flexible polymers because they become soft with heating and harder after cooling. They usually have the simplest formulae, e.g. polystyrene, polycarbonate or polyethylene.

Thermosets

This group of polymers has a much more complex structure: the chains bond covalently (by sharing electrons) to adjacent chains, often called *crosslinking* (Fig. 2.6). This is the strongest form of chemical bonding and so thermosets are rigid and resistant to deformation from mechanical stress or heating. If the crosslinking formed during polymerization is excessive the polymer can become brittle and liable to fracture. This group is also the most resistant to organic solvents.

Elastomers

The chief characteristic of this group is that the polymer chains are extendable (Fig. 2.7). After an

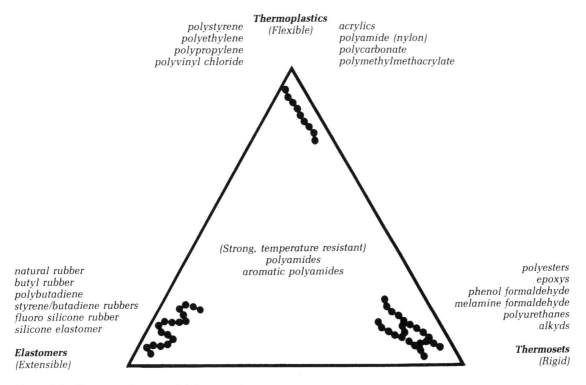

polystyrene
polyethylene
polypropylene
polyvinyl chloride

Thermoplastics
(Flexible)

acrylics
polyamide (nylon)
polycarbonate
polymethylmethacrylate

(Strong, temperature resistant)
polyamides
aromatic polyamides

natural rubber
butyl rubber
polybutadiene
styrene/butadiene rubbers
fluoro silicone rubber
silicone elastomer

polyesters
epoxys
phenol formaldehyde
melamine formaldehyde
polyurethanes
alkyds

Elastomers
(Extensible)

Thermosets
(Rigid)

Figure 2.4   Common polymers and their categories.

Figure 2.5   Thermoplastics. Linear individual chains.

Figure 2.6   Thermosets. Crosslinked polymers.

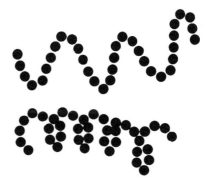

Figure 2.7 Elastomers. Coiled chains. These are in a state of continual movement with rotation around the carbon backbone.

imposed stress is released there is an element of mechanical recovery. Due to the rotating arrangement of the bonds the chains are often coiled which gives elastomeric molecules great potential for extension under applied stress. The chains can retract dramatically after the release of any stress and exhibit an ability to recover and re-orientate themselves into their original position: a form of plastic memory. This can only happen when the material is above T_g (the glass transition temperature). Elastomers can be thermoplastics or thermosets. If we want a more rigid but 'rubbery' material a crosslinked (thermoset) elastomer would be chosen. The rubber of ordinary car tyres fall into this category. The cross-linking of rubber by sulphur is known as *vulcanization*.

Application

The polymers we use are usually applied to a surface as a liquid, and after application, we expect them to harden. We also expect this hardening mechanism to be permanent at ambient temperatures but the degree of solidity needed depends on the conditions of use. For example we may require the film to retain sufficient flexibility to move with the substrate. This is particularly true for timber substrates which may vary by as much as 4% extension over one season. In certain situations a film that has achieved too great a degree of hardness will be unable to flex, be less likely to resist stress and become brittle and fracture. The chemistry of the polymers determines our expectation of the performance of paint films, which is of course also temperature dependent.

2.2 Glossary

Alkyds   These are a class of branch-chain polyesters and form an insoluble thermosetting polymer film, formed by condensation polymerization.

Acrylics   Polymethylmethacrylate, PMMA, is a thermoplastic. In a glassy sheet form it is called *perspex*).

Amino resins   (Urea-formaldehyde and melamine-formaldehyde)   These resins give good hard glossy films with good adhesion to metals. As they are insoluble in common solvents they must be diluted with butanol or the higher alcohols, or with esters for application. They can be converted to crosslinked thermosets either by adding acids or by being exposed to high temperatures, 110–170 °C. They are often combined with alkyds to improve water resistance.

Condensation polymerization   The reaction of compounds to form long-chain molecules (polymers) with water as the usual by-product.

Co-polymers   Polymers formed from two or more different monomers. The polymer chains contain a mixture of monomers and exhibit properties different from those of the constituent polymers.

Crystalline polymers   The principle of crystalline polymers is illustrated in Fig. 2.8. When individual chains of a polymer align and form a repeating pattern, they are said to be *crystalline*.

Emulsion   The suspension of one liquid within another. As 'emulsion paints' carry suspended solids, i.e. pigment resins in a liquid medium, they should be more correctly classified as a 'latex' solution, which is a milky fluid with substances in suspension.

Epoxy resins   Ethylene oxide (epoxide group) R'-CH-CH$_2$O. These are very strong resins and can adhere well to many surfaces including metals. They show great stability as they start to harden, and exhibit high polarity which assists in local secondary bonding. Epoxyresins are linear polymers which are always used with another component to act as a crosslinking agent to complete a typical crosslinked thermoset structure. One drawback is that many of the secondary agents used are highly toxic. These compounds can be cured at very high temperatures (180–200 degrees °C) with phenolic resins.

Esters   These are compounds formed by the reaction of fatty acids (with -COOH group) with alcohols. The polymer molecules (polyesters) can also crosslink as they solidify and hence fall into the category of thermosets.

Gel   The suspension of a solid within a liquid.

Glass-transition temperature   Often expressed as T_g this is the temperature at which a material changes from the liquid to an amorphous solid. This is the definition of a glass. Most paint films do not become glass-like but stay partly in the liquid state and partly in the solid state. If the entire film became a glass it might crack.

Ketones   These have a carbonyl group attached to two hydrocarbon groups (-CO-). Ketones are highly reactive. The simplest one is acetone (propanone).

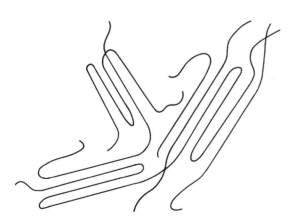

Figure 2.8   Crystalline polymers.

Lacquers From the ASTM D16:1991[1] glossary of paint terms a lacquer is 'a coating composition that is based on a synthetic, thermosplastic, film-forming material dissolved in organic solvent that dries primarily by solvent evaporation.' Typical lacquers include those based on nitrocellulose, other cellulose derivatives, vinyl resins, and acrylic resins. These films can deteriorate in the presence of household chemicals and surprisingly, of human perspiration. One major problem is that plasticizers can migrate from vinyl fabrics into lacquer films and cause softening.

Micron A unit of measurement. A millionth of a metre, i.e. 1×10^{-6} m. There are 1000 microns in a millimetre. Paint films are commonly of the order of hundreds of microns in thickness, i.e. 1×10^{-4} m.

Microporous This is a popular term given to some paint films today which have been engineered to give greater permeability. All materials, including paint films, are able to absorb gases and to some extent liquids. The term was originally used to describe fabric designed to keep out liquid-phase water molecules but allow in water the gas phase. The pore size can be altered but care should be taken to ensure that films retain the properties required for the application.

There are tests for determining the permeability of films. This is done chiefly by measuring physically the transfer of water through a sealed container as in ASTM D1653:1991.[2] They state that water vapour transmission rate is not necessarily a linear function of temperature and relative humidity and that there is not necessarily a proportional relationship between film thickness and its effectiveness.

Oils Oils are neutral liquid hydrocarbons (i.e. neither acid or alkaline on the pH scale) and come in three classes.

- *Fatty Oils* or *fixed oils* are organic compounds from animal or vegetable sources and comprise polyglycerides or esters of fatty acids.
- *Mineral oils* (hydrocarbons with a higher molecular weight) come from petroleum and coal sources.
- *Essential oils* are volatile hydrocarbons, which have characteristic odours and are extracted from plants.

Paints, especially traditional ones, are often quoted as having a particular *oil-length*. This relates to resin-oil ratios as shown in Table 2.1.

Paint A finished coating to a material that consists of the suspension of particles in a liquid phase that subsequently hardens to form a solid film.

Table 2.1 Resin-oil ratio terminology

Resin/oil ratio	Terminology for varnish/paint
$1:\frac{1}{2}$–1:2	Short oil
$1:2$–1:3	Medium oil
1:4–1:5	Long oil

Plasticizers Plasticizers work by depressing the glass-transition temperature of polymers, weakening the bonds between individual polymer chains, making them softer at ambient temperatures and causing a greater tendency to flow (lower viscosity) with a higher permeability. They are not ordinary solvents (which would evaporate) although they are dissolved in the polymeric material. They can also make the polymer less durable in the long term.

Polyurethanes These are saturated polyesters cured by polyisocyanates to produce urethane coatings. *Warning*: isocyanates have a high vapour pressure and can affect the lungs giving asthmatic/bronchial symptoms. Polyurethanes form hard films in a short time with crosslinked structures characteristic of thermosets. They can be degraded by silicones which destroy adhesion by affecting surface-wetting characteristics.

Resin This term is used to describe a synthetic plastic material. Resins are usually made up of acids and the initial compounds of polymeric materials in a highly polymerized state.

Shellac This substance is made from the secretions of tiny insects (*Laccifer lacca*). The insects are eventually covered in a hard resinous substance making them almost twiglike. The insects are collected, ground down, melted and compressed into sheets. It takes 150,000 creatures to make 550 grams of shellac and this enormous wastage inspired Leo Bakeland to find a synthetic substitute (Bakelite).

Thixotropy A substance is thixotropic when it is reduced in body and volume after the application of stress. It requires a shearing force to break down the molecular structure to make it flow like a liquid. Thixotropic paint will spread from the tin without dripping. If the shearing force is removed, the paint will return to a thixotropic gel.

Toughness This is a term used by a materials scientist with a specific meaning: it is the ability of a material to resist the propagation of a crack.

Varnishes Varnishes are clear finishes (sometimes called lacquers) which have the same composition as paint films but lack the pigmentation. As they do not have extenders and pigments, they are pure forms of normal paint compositions and so will harden faster and be more highly viscous, which can be a problem in application. ASTM D154:1989[3] states: 'Most varnishes are predominantly yellow but the colour of the liquid varnish is only a preliminary indicator of the colour of the dried varnish film. The initial colour may bleach or darken depending on the conditions of exposure.' The viscosity of the liquid is important to allow for satisfactory brush application. As these films harden by oxidative polymerisation a skin may form on top of the liquid in a can which is then insoluble in the rest of the liquid.

Vinyls A vinyl group is the simplest structure that can be used to obtain a polymer chain. If the most basic monomer building block (ethylene, $CH_2=CH_2$) has one (or more) of the ethylene hydrogens replaced with another atom or group such as chlorine, fluorine or benzene, a new monomer called 'vinyl chloride' is obtained. The following are examples of vinyl monomers:

-Cl Vinyl chloride
-F Vinyl fluoride
-C_6H_6 Vinyl benzene or styrene
-$OCOCH_3$ Vinyl acetate
-$CH_2=C(CH_3)COOH$ Methylmethacrylate

Vinyl is derived from the latin vine indicating both sides of a branch. The carbon backbone resembles a branch when the individual monomers are polymerised. 'Vinyl' is often loosely used as a term for 'sheeting'. A vinyl plastic is a thermoplastic and so would be likely to have flexural properties.

Viscosity Viscosity is the resistance to flow possessed by a liquid. Paint films when set should be highly viscous. Ordinary glass (sodium silicate) is highly viscous but over time can be seen to 'creep'. If a paint film has too low a viscosity then it will 'run' after hardening and will be defective.

2.3 Adhesion

Introduction

Adhesion is an area of materials technology which used to be left out of formal building textbooks. This is due to the comparatively recent development of the subject as a major discipline in its own right.

Originally adhesion relied on the correct positioning of materials so that they would not be displaced by weak forces. Many timber adhesives were used to position firmly joints which they owed their mechanical strength to the physical lapping and securing of one piece of timber to another. In the twentieth century glues have become an integral part of a jointing system and materials are true *composites* (with a combined performance which is superior to the individual components).

Modern glues are now as strong or even stronger than the materials to be glued together. The mechanics of this technology can also be calculated but builders require a more specialized knowledge to ensure that materials to be jointed are correctly prepared and are compatible with the jointing methods used.

Adhesion is the result of several factors. The simple notion of 'stickiness' is not specific enough to understand the subject. The mechanics of adhesion are just as applicable to paint technology, and also to the fixing of larger-scale materials such as rendering or tiles. For convenience we will consider adhesives of two types: *organic* adhesives are polymeric and used for jointing a wide range of materials; *inorganic* adhesives are used to fix ceramic materials. In the next section we consider the parameters essential for adhesion to take place.

Factors enabling adhesion to take place

Mechanical adhesion Good mechanical adhesion means that one material resists physically being pulled out by another. A surface topography that gives a dovetailing effect is ideal. This profiled detail is known as *undercutting* and to be successful the inclines must be at an angle of at least five degrees (Fig. 2.9). In reality undercutting of surfaces rarely occurs, and although the surface topography may be rough, its appearance is more likely to resemble a miniature forest landscape with a pyramidal profile that is 'overcut'.

Although this surface landscape is obviously not ideal, the depth of surface roughness creates a porous surface that can assist in giving greater penetration of the adhesive. Surface roughness may not necessarily be needed and good adhesion can be obtained on perfectly smooth surfaces if other factors are adequate.

Diffusion At the microscopic level, the molecules in a solid are constantly vibrating or even swapping places with other molecules. Thus molecules of an adhesive on a surface can move a small distance into an adjacent surface. The notion of *migration* is

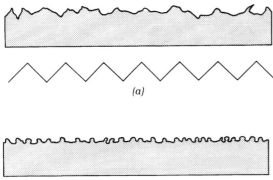

Figure 2.9 Surface topographies (real and idealized): (a) a pyramidal profile that is 'overcut'; (b) cold metal without surface finishing – the crystal structure gives a degree of undercutting.

becoming better understood, particularly from adhesive or sealing compounds into other materials. For example after 100 hours molecular segments of a polyisobutylene travelled from one sheet to another a distance of 10 microns. (A J Kinloch *Adhesion and Adhesives*) This is a small distance but in adhesion terms if one material can diffuse even a distance of 1–2 nanometres (10^{-9} m) there can be a theoretical increase of between five and nine times in the strength of adhesion. With diffusion there can also be a more direct interaction between long chain molecules: they can 'entangle' giving a physical meshing of material on a very small scale. This process can be accelerated by the use of liquids that are called *solvent welders*. These compounds really accelerate the rate of diffusion of one material into another.

Electrical attraction   Electrostatic forces are strong forces which result from electron transfer. Adhesive and substrate may have very different structures with different electronic configurations, and charge may be transferred from one to the other causing electrostatic attraction.

Adsorption   Adsorption is the most important mechanism for adhesion and is different from diffusion. It relies on intermolecular bonding. The most common form of bonding is due to van der Waals' forces, which are weak *secondary bonds*. Ideally an adhesive will cause *primary bonding* between the surfaces. Primary bonds are chemically strong bonds which include *covalent, ionic* and

metallic bonding. These are the basic bonds that form the internal atomic structure of materials, so this type of adhesive will have a strength comparable to, or greater than, the materials to be joined.

Having considered the mechanism of adhesion, we will now examine the important characteristics that the adhesive must have for adhesion to take place. These relate to the liquid character of the adhesive and its ability to 'wet' a surface adequately.

Surface-wetting characteristics and surface conditions

Low contact angle   A liquid used for adhesion must have a *low contact angle* with the surface of the substrate (Fig. 2.10). The closer a liquid gets to being spherical the more the contact with the surface is tangential and gives a high contact angle. This principle is used to produce surfaces that are deliberately only slightly 'sticky', and a good example is the peel off strips on message pads which have a surface that shows miniature spheroids with a limited contact.

Low viscosity   The liquid used for adhesion must have a *low viscosity* for *surface wetting* to take place. Polymeric compounds with a low molecular weight are generally less viscous.

No air   All *air must be eliminated* completely from the contact surfaces to be joined. Sometimes vacuum

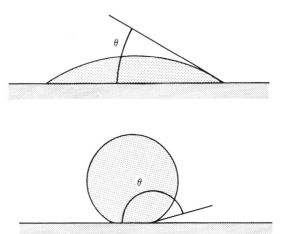

Figure 2.10 Adhesion and surface wetting. Good adhesion requires good surface wetting characteristics achieved by a low contact angle (θ). A high contact angle will give low adhesion and peel off characteristics as exploited in re-usable spray mount or yellow message stickers.

methods are used to eliminate all air from the material. There is a direct relationship between porosity and strength. In an experiment measuring the strength of aluminium/epoxy joints removing air increased the strength of the joint by 30%.

Cleanliness   The *cleanliness of surfaces* to be joined is critical. Impurities can absorb water vapour and contaminents can affect surface spreading and interfacial contact.

Surface preparation and surface wetting

There are forces present in an adhesive situation which are calculable and which result from the movement of molecules within the adhesive film. One of the problems in fully understanding these films is that often the adhesive mechanism is a dynamic one, and various molecules are in motion. This rate of movement normally slows down as full adhesion develops. The actual rate of movement is partly due to the power of attraction by molecules within the body of the materials to the surface layer. This force of attraction is due to van der Waals' forces, and the force of attraction is sufficient to displace molecules to the side. As molecules are displaced and travel a discrete distance they do *work* and work can be calculable in terms of energy. Whenever phenomena like these can be evaluated in energy terms *thermodynamic equations* can be used to model the situation.

Thermodynamics can explain when materials start to change their state of being. They change from one *state* or one *phase* to another, i.e. from liquid to solid to gaseous. Although adhesion requires surface wetting to take place, surface wetting also depends on the state of the surface free energy values of these different phase states, and, for spontaneous wetting to occur, the surface free energy value of the solid–vapour interface must equal or be greater than the sum of the surface free energies of the solid–liquid interface and the liquid–vapour interface. Alternatively, the surface free energy of the solid must be equal to or greater than the sum of the solid–liquid interface, the liquid–vapour interface and the equilibrium spreading pressure. All of these parameters can be quantified and so a prediction given for surface wetting.

The condition of the surface is critical for surface wetting to take place. It is often an advantage to pre-treat the surface with a primer. The function of a primer is to modify the surface of the substrate by:

- removing weak boundary layers, e.g. oil and grease, or even plasticizers which might have migrated to the surface and become contaminants
- maximizing molecular contact by increasing the surface free energy
- introducing strong primary bonds. The primer should form a layer that has dual compatibility with two or three materials, i.e. adhesive and substrates, whereas the adhesive might have an optimum strength with only one material, i.e. the primer
- changing the surface topography. This could mean slightly eroding the surface to create a deeper topography, greater porosity and hence greater depth of penetration
- assisting in hardening (possibly by initiating a reaction with the atmosphere that provides a more stable coating and inhibits migration of compounds from the substrate)
- protecting the substrate prior to bonding

Although the knowledge needed for predicting accurately adhesive behaviour is far beyond that required by a practising architect or designer, it is important that an appreciation of the complexity of the subject is grasped, so that careful and informed decisions can be made about all materials and components used for bonding.

Some concepts of the theory are difficult to grasp but the following examples illustrate some of the points covered.

Flooring case study

Thermoplastic tiles had been laid directly on boarding using an unknown adhesive. In taking up the flooring it was noticed that the tile material was in fact very brittle and fractured easily, with cleavage around boundary surfaces of large-scale particles in the flooring material. The adhesive was very 'sticky' i.e. it was in a *liquid/rubbery* phase and, as the flooring material was lifted, strands of the adhesive elongated and adhered to adjacent newspaper.

The two items of interest here are the brittleness of the thermoplastic tile which had changed from being a flexible sheet, and the unexpected almost liquid state of the adhesive after a long period (Fig. 2.11).

With many materials there is a tendency for 'like to dissolve like' and it appears that the plasticizer had migrated from the body of the tile into the adhesive and was helping to keep it liquid.

Adhesive tape study

A small portable steel cupboard which contained bottles of methylated spirit and white spirit for many months also contained some cardboard wrapped with

Figure 2.11 Embrittlement of vinyl (thermoplastic) tiling. The plasticizer migrates into the flooring, keeping it 'sticky' years after the initial laying. This photograph shows the high reflectance from the still 'wet' surface, the pattern of embrittlement and the concentration at the edge of the brittle cracking, with crack diversion around particulate material.

brown adhesive tape. On unwrapping the cardboard it was obvious that the tape had very uneven patches of adhesive, with some areas that had become almost transparent. Where the brown adhesive coating had almost completely disappeared, any remaining was in a liquified state. The smell from the solvents contained in the cupboard was quite powerful.

The concentration of chemicals in the gaseous phase was sufficiently great to change the surface wetting characteristics of the adhesive. It affected the stability of the adhesive film to such an extent that liquification had taken place and partial change of the adhesive into a gaseous phase may also have occurred (Fig. 2.12).

Reactions between materials are not usually so dramatic or happen so quickly. Sometimes gaseous phases present can affect the stability of flooring and roofing materials, particularly those that require intermediate bonding that is totally reliant on adhesion.

The chief points to remember are:

- There must be *compatibility* between the adhesive and the two substrates to be joined.
- Layers must be *isolated* in situations where the constituents of a particular substrate are not known. New layers should not be indiscriminately bonded on top of each other but a primer should be used to isolate the surface of the questionable substrate.

Figure 2.13 shows the consequences of poor adhesion, when these principles have not been followed.

2.4 Degradation of thin surface films

As paint films whether external or internal are the first layer of the building fabric to be exposed to the elements, they are subject to wildly fluctuating

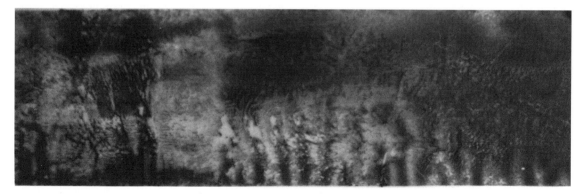

Figure 2.12 Adhesive tape. After reacting with vapours from white spirit and methylated spirit the brown adhesive layer has de-stabilized and changed into a liquid phase, revealing the clear tape below.

Figure 2.13   Adhesion between two pieces of blockboard. *Left* and *middle*: The One area where there has been true adhesion between the two faces – approximately 65 mm × 40 mm on an area 1200 mm × 1200 mm. Adhesive had been applied to one surface and then the two surfaces had been screwed together. The two panels were separated four years later, parting with a loud crack (strong bonds were broken in the cellulose tubes and energy was released as sound). *Right*: The transfer of adhesive from one side to another, showing minimal surface wetting. The moral of applying adhesive to both faces is clear.

temperatures and conditions. The following criteria of performance should be used as a checklist for film quality.

Mechanical wear and abrasion   The film should be hard but not brittle. It should of course be tough and it may be necessary to evaluate its hardness by impact testing.

Fracture development during expansion   Some substrates (especially timber) have expansion cycles that must be tolerated by the film. Look for *flexible films*, and, if critical, find out their maximum extensibility.

Water absorption   Most polymers do absorb water but the degree to which they do this varies enormously. For example, nylon can show significant swelling. When polymers absorb water their structure is stiffened making them less resistant to bending and this in turn can promote the generation of hairline cracking under stress. Water absorption can also affect the glass transition temperature of a polymer by lowering this critical temperature. This can cause the polymer to move further into the liquid phase and show permanent deformation under stress in warm conditions. Look for *microporous* paints which can assist the movement of water vapour across the film without absorption.

Degradation under ultra-violet light   Polymers in paint films can become brittle after stress cracking if they undergo excessive crosslinking due to the effects of energy bombardment from ultra-violet light. Look for paints containing *ultra-violet light stabilizers.* These are compounds that absorb harmful UV and radiate this energy as heat. Obviously, it is advisable *not* to paint in strong sunlight: the crosslinking reactions can cause variations in the depth of the applied coating as uneven reactions take place.

Defects in paint films

Having considered the technology of polymers in general we will turn to the subject of the defects. Some major defects are considered here prior to more detailed information on surface coatings. Reference should be made to BS 3900, particularly section H on the evaluation of paint and varnish defects. The American standards (denoted as ASTMs in the text) are quoted here as they are useful for their clear definitions.

Blistering   Blistering can occur for a number of reasons. One is the presence of moisture in the substrate which cannot migrate successfully through the finished film (Fig. 2.14). Solvents can also be trapped by rapid surface hardening and reduce 'wet edge' time.

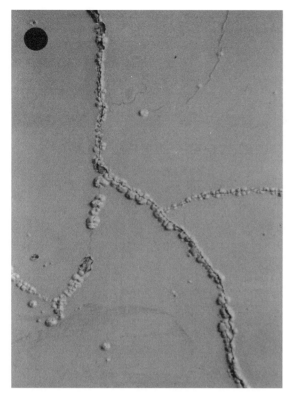

Figure 2.14 Blistering. In this example, paint failure is occurring along the lines of map cracking in a plaster substrate. Moisture in the structure behind migrates through these cracks, and expands in the warmer internal environment. It cannot escape below a relatively impermeable film, causes blistering, which increases in size until the maximum extensibility of the film is reached. The film then fails.

Chalking Chalking is defined in ASTM D4214: 1989[4] as: 'the formation on a pigmented coating of a friable powder evolved from the film itself or just below the surface'. In Part H6 of BS 3900 it is a 'loosely adherent fine powder on the surface of a paint coating, arising from the degradation of one or more of its constituents'. It is a common type of degradation on exterior paints. Paint has constituents which act as fillers, so that if the holding vehicle breaks down, pieces of filler as well as pigmentation particles will be released.

Checking Checking is not a term used in BS 3900 which talks in terms of degrees of cracking but some people do use it. In ASTM D660:1987[5] checking is the 'phenomenon manifested in paint films that do not penetrate to the underlying surface. The break should be called a crack if the underlying surface is visible.' Fig. 2.15 shows a typical example.

Cracking In BS 3900 Part H4 cracks are rated from 0 to 5 ranging from not visible to large cracks over 1 mm in width. In ASTM D661:1986 cracking is 'that phenomenon manifested in paint films by a break extending through to the surface painted' (Figs 2.16 and 2.17).

Erosion In ASTM D669:1981[6] erosion is defined as: 'the phenomenon manifested in paint films by the wearing away of the finish to expose the substrate or undercoat. The degree of failure is dependent on the amount of substrate or undercoat visible.'

Efflorescence In ASTM D1736:1984[7] efflorescence is defined as: 'A condition that occurs when soluble salts in the dried paint film migrate to the film surface during exposure. Efflorescence is seen as either a light, medium or heavy deposit of crystals.' It often occurs in cycles of low temperature and high humidity with condensation forming on the coating surface. Such a cycle is used to test paints for efflorescence.

Figure 2.15 Checking and cracking. On this wooden garage door there is checking and then cracking parallel to the grain showing lack of adhesion in line with grain detail. This could indicate an aged timber surface that should have been abraded or planed to remove surface defects, or filled and sanded to give a smooth and homogeneous background.

Table 2.2 Defects in paintwork and remedial treatment

Defect	Typical causes	Remedial treatment
Adhesion failure	Application to damp or dirty substrates; or subsequent entry of moisture, e.g. through open joints in woodwork	Flaking, peeling or poorly adhering material should be removed. Where moisture is the cause, it should be ensured that the substrate is dry before repainting
	Failure to prepare or pretreat non-ferrous metals correctly	Defective material should be removed as above
	Omission of primer or use of unsuitable primer	Defective material should be removed as above. Refer to appropriate substrate clause in section on priming
	Application to powdery or friable substrates	Defective material should be removed as above. Application of a penetrating primer or sealer may be necessary
	Application to hard, dense substrates, e.g. glass or glazed surfaces	Defective material should be removed as above
	Apparent loss of adhesion on iron and steel, may be due to detachment of millscale	See below
Blistering	Blistering is usually indicative of liquid or vapour beneath the coating. The presence of water is a frequent cause. On woodwork, resinous material may be responsible	Depending upon the extent and severity of blistering, preparation may be confined to removal of isolated blisters or complete stripping may be necessary. Where moisture is the cause, time should be allowed for drying out. Blistering on resinous external woodwork may be influenced by choice of finishing colour
Chalking, powdering	Slow erosion and chalking on lengthy exposure, expecially externally, is a characteristic of many paints and wood finishes. It is not usually regarded as a defect unless it occurs prematurely and profusely, when the causes may be as follows: ● conditions of exposure exceptionally severe ● earlier coats in system have failed to satisfy porosity of substrate ● incorrect or unsuitable formulation	In absence of other defects, lightly chalking surfaces may require only washing and light abrasion to provide a satisfactory base for further coats. Heavily chalked or powdery surfaces will require more vigorous cleaning or abrasion combined if necessary with application of a penetrating primer
Colour defects, e.g. fading, staining, 'bleeding', or other forms of discolouration	Some loss of colour of paint may occur on lengthy exposure to bright sunlight but is not usually significant. Early loss of colour may be due to use in unsuitable conditions, e.g. external use of a colour intended only for interior use. Chemical attack may cause change or loss of colour	If necessary, consult manufacturers regarding selection of colours or types of finish for repainting
	Oil-based finishes tend to yellow in situations where direct daylight is excluded. This is more obvious with white and light-coloured finishes	Yellowing is not usually sufficiently marked to be significant. If freedom from yellowing is important, consult manufacturers for guidance on selection of oil-free coatings
	Apparent colour change may be due to masking of colour by surface chalking (see above), to efflorescence especially on external rendering, or, on external plywood treated with wood stain, to diffusion of water-soluble salts contained in adhesives	Normal cleaning usually removes surface deposits. Efflorescence and diffusion of salts on plywood may recur until source is exhausted

continued

Table 2.2 continued

Defect	Typical causes	Remedial treatment
	Failure of clear finishes on external woodwork may result in discolouration of exposed wood	Clear finish should be removed completely. Sanding or scraping may remove discolouration, but application of coloured wood stain may be necessary to achieve uniform appearance
	Constituents of the substrate or previous coatings can cause discolouration	Wash with detergent, use alkali resisting primer and re-paint
Cracking, other than that due to structural movement	Cracking is usually indicative of stresses within the coating film, caused, for example, by applying hard-drying coatings over soft coatings. It may also be the initial stage in adhesion failure (see above). Cracks may be confined to the finishing coat or extend through the thickness of the film	If cracking is slight and confined to the finishing coat, rubbing down may provide a satisfactory base for recoating. If cracking is severe or extends through the thickness of the film, complete removal may be necessary
Damage to coating	Mechanical damage, e.g. by abrasion, impact or vigorous cleaning

Graffiti | Where surfaces are subject to hard wear, specialist coatings may be required. Consideration should also be given to the use of wear-resistant materials, e.g. ceramic tiles or plastics, where practicable |
Gloss, loss of	Some loss of gloss after lengthy exposure, especially externally, is to be expected and may be the first stage in chalking. Where it occurs prematurely, possible causes are as described for premature chalking (above)	Loss of gloss in the absence of other defects is not usually significant in relation to maintenance treatment
Millscale, detachment from painting iron and steel	Poor initial preparation leading to formation of oxide film below	Removal of millscale, e.g. by blast-cleaning or flame cleaning, may be impracticable as a maintenance operation and is costly, hence the desirability of effective initial preparation. There may be no alternative to manual cleaning to remove millscale as it loosens, but this may extend over several repaints
Organic growths, e.g. moulds, algae, lichen, moss	Micro-climate, poor maintenance	Consider modifications to design or environment which may eliminate or reduce causes of failure
Rust-spotting or rust-staining on painted iron and steel	This usually indicates that the thickness of the paint system is insufficient to provide protection on peaks and edges. It may result from application of an inadequate system initially or at the last repaint or from erosion of the film during exposure. A further possible cause is failure to use a rust-inhibitive primer	Depending upon the severity and extent of the defect, treatment may range from manual cleaning and priming of localized areas to overall removal of the coating and treatment as for new iron and steel. Consideration should be given to increasing the film thickness of the system or to reducing the intervals between repaints until an adequate thickness has been built up

Note: 'Paintwork' refers generally to paints, clear finishes and wood stains.
Source: Based on Table 18 BS 6150:1991

Figure 2.16   Cracking. Paint on a wooden door, Smithfield Market, London. The paint coating was sufficiently thick to develop major cracking in line with the grain (known as preferential cracking according to the structure of the substrate), after ageing of the film and subsequent embrittlement.

Flaking   In ASTM D772:1986 flaking is 'the phenomenon manifested in paint films by the actual detachment of pieces of film, either from its substrate, or from paint previously applied'. Flaking or 'scaling' is generally preceded by cracking or checking or blistering, and is the result of loss of adhesion usually due to stress-strain factors.

More details of the causes and treatment of defects in paintwork are given in Table 2.2.

Figure 2.17   Paint as a sheet material. Paint has been applied to a rubber dustbin lid. Due to poor adhesion with the subtrate the film has hardened, lifted and then fractured in a brittle manner. The angular cracking patterns are similar to those in ceramics, and the plastic has hardened to a glassy state.

Notes

1   *Terminology relating to paint, varnish, lacquer and related products.*
2   *Test for water vapour transmission of organic coating films.*
3   Methods of testing varnishes.
4   Method for evaluating degree of chalking for exterior paints.
5   Method for evaluating degree of checking for exterior paints.
6   See also Exterior paints D662:1986 and Interior paints D4213:1992.
7   Test method for efflorescence of interior wall paints.

3 Applications of polymeric materials

3.1 Surface coatings technology

Finishing coatings perform protective and decorative functions. Unfortunately, the protective function is often underestimated: coatings must have the ability to resist a great range of external stresses, whether physical, chemical or biological, and must stay stable over a wide range of temperatures. This demands a high level of engineering on the part of the paint technologist. It is also important that the coatings used are specified for the particular conditions present and that a specification covers adequately the correct preparation of the substrate. To ignore the condition of the material to be coated is effectively to build in failure of the film and an unacceptably short life. As coatings are now so specialized it is important not to use on site substitutes that do not meet the original specification, and it is important not to mix paint systems. Where transfer of moisture must be maintained across the structure, the coating system must have the right porosity.

All polymer-based materials use solvents for application. As part of a broad commitment to improve air quality the European Community is pledged to reduce VOC from the storage and distribution of petrol emissions (volatile organic compound) 90% over a 10–15 year period. This is to be followed by controlling emissions from car refuelling and then from general solvent emissions. VOC's are thought to be involved in the formation of photochemical oxidants which affect human health, all vegetation and contribute to global warming as greenouse gases.

The total emission of VOC's in the EC is 10×10^6 tonnes per annum. Table 3.1 shows their origins.

Legislation to date against VOC's has been directed to improving air quality. Further legislation will have the effect of further limiting their use due to toxicity: many VOC's are carcinogenic, mutagenic or teratogenic (having toxic effects on reproduction). EC legislation is currently being drafted to increase awareness of the existing directives relating to *Classification, packaging and labelling of dangerous preparations.*[1] The legislation will clearly identify which compounds are dangerous to health and to the environment in terms of hazards to aquatic and non-aquatic life. Safe disposal of these preparations is also an issue. Directive 77/728 relating to the *classification, packaging and labelling of paints, varnishes, printing inks, adhesives and similar products* is also relevant. Manufacturers have to adapt quite rapidly to these directives and EC legislation is affecting the drafting of British Standards.

The Environmental Protection Act of 1990 has been followed by EC legislation and new British Standards. These new standards deal with aspects of health and energy. They require the gradual phasing out of solvent-based products and much better ventilation and protection for workers. Employers must provide proper ventilators, not just masks for workers. The new COSHH regulations require that when solvents are used in preparations (including paint coatings, resin floor finishes and preservatives) workers are properly protected. The new legislation also requires the identification of hazards in the construction industry. The effects of this legislation are three-fold. They

Table 3.1 Composition of VOC emissions within the EC

Source	Percentage contribution
Motor vehicles	35–40%
Solvents	40%
Storage and distribution of petrol	5%
Refuelling of cars	2%
Industrial production	Remainder (13–18%)

ensure a steady reduction in the use of toxic and harmful substances by insisting on lower concentration levels, the minimization of VOC's which deplete ozone in the atmosphere (contributing to global warming), and there is an energy saving in the processing of these compounds.

Currently the 1990 Environmental Protection Act, covers pollution from transport and power plants not the local or long-term use of solvents from paints. Such use is a factor in measured air quality and will be controlled under new EC legislation. The UK paint industry produces 170×10^6 litres of decorative paint per annum.

As a result of legislation and growing public awareness of these issues more water-based systems for coatings and better water-based adhesives and binders have been developed and produced commercially.

The major source of reference for specifiers is the Code of Practice for *Painting of Buildings* BS 6150:1991. This document covers the design and specification of coatings, the materials used, coating systems, application, maintenance and the 'organisation of work with new environmental considerations which are linked to health and safety' (clause 24). Although inspection, sampling and testing is included as a section in this Standard, it is worth making reference to BS 3900 and to ASTM series of standards for detailed descriptions of possible tests. BS 3900 has approximately 80 parts (up to 1993) which cover all possible tests needed to establish the quality of films. These are grouped into alphabetical series. Group E gives the mechanical tests on paint films. Group F gives the standards for durability, Group G the environmental tests and Group H the evaluation of paint and varnish defects. The ASTM standards are very readable and clear in their definitions, and include the science behind the technology of paints. One of the most important tests is establishing the moisture content of the substrate. Incorrect moisture content is the source of many coatings failures. A great deal of time and effort is often spent establishing the cause of failures in films (with attempts at the apportionment of responsibility). These failures could have been avoided by careful inspection prior to and during coating application. (See Table 2.2, section 2.4, which summarizes the main defects and possible remedial action.)

Characteristics of paint surfaces

Coatings, surface texture and resultant colour   The overall colour and opacity of a surface coating depends on the properties of the pigments, not just their body

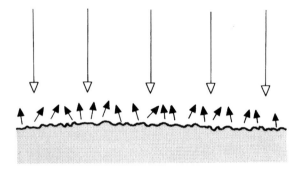

Figure 3.1   Rough surfaces showing scattering of light.

colour, i.e. the selective wavelengths of light that they do not absorb. The actual size of the particle of pigment and how it physically distributes light are equally important. Textures of surfaces can be divided in rough and smooth.

Rough   These textures gives a greater vibrancy to colour. The uneven surface reflectances give a wide range of tones, e.g. hand-made bricks (Fig. 3.1).

Smooth   There are two categories of smooth surfaces.

* *Hard* surfaces have a greater surface reflectance and light is bounced back with only limited absorbency by the substrate, e.g. marble and glass exhibit a mirror-like quality (Fig. 3.2).
* *Soft* surfaces have small scale surface irregularities which allow for a high degree of absorbency, and great intensity of colour is given to the material with very little reflection. The best examples of this surface texture are soft fruits or flowers (Fig. 3.3).

The importance of surface texture to the classification of colour is best understood by a comparison of artists' paints. Cheap paints yield a watery solution with

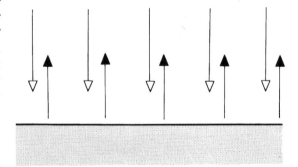

Figure 3.2   Smooth but hard surface showing complete reflectance of light and no absorbency.

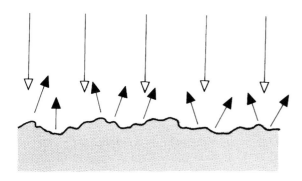

Figure 3.3 Smooth but soft surface showing part absorption and a diminished reflectance of light.

disappointing covering power and have a noticeably 'gritty' appearance. More expensive paints are completely smooth in appearance and have a great covering power. The particle size of the pigment is far smaller and there are fewer impurities.

Smaller particles of pigment give a greater density of colour and hence covering power, because they are able to scatter light internally with a minimum of reflection. There is no need for paint films to be 'thick' to be effective. The science of determining the hiding power of a paint by the physical qualities of the pigmentation is known as *reflectometry* and ASTM standard D2805:1988[2] estimates relative efficiency.

The colour of paints have been determined historically by the pigments available. Some colours, such as red or purple, were extremely rare due to a limited source of supply. They were only available to the heads of social hierarchies. Colours in the green, dark blue or brown ranges were readily available from a great variety of natural pigments for ordinary usage before the First World War.

Traditional pigment sources (Rivington *Notes on Building Construction*: Paints and Varnishes (1901))

Blacks Soot from oil, coal, resinous woods. Also bone black.

Blues Prussiate of potash from the remains of old leather, blood, hoofs and other animal matter boiled with iron fillings. Indigo is obtained from plants from Asia and America. Ultramarine from lapis lazuli, cobalt blue from an oxide of cobalt and blue ochre from a natural clay.

Browns Oxides of iron. Umber is a naturally occurring clay stained with oxides of iron, and may be burnt to produce darker browns. Also sienna and burnt sienna. Also Brown ochre.

Yellows Chromates of lead. Dilute solutions of lead acetate or nitrates. Bichromate of potash. Yellow ochre is a natural clay stained by iron oxides. Yellow lake is made from tumeric. Alum, etc.

Red Red lead and vermilion (a naturally occurring sulphide of mercury known as Cinnabar from China). Vermilion was also made artificially. Indian red, a naturally occurring iron ore from Bengal. Venetian red was obtained by heating iron sulphate (a waste product from tin and copper works).

Greens Copper compounds, arsenic, etc.

Sources of colour from vegetables (limited)

Red	Madder root
Orange	Onion bulb
Yellow	Saffron
Dark Brown	Walnut fruit

Rare sources of colour

Red	Cochineal insect (Carmine)
Blue	Azurite (mineral)
Green	Malachite (mineral)
Purple	Murex (mollusc)

Modern pigment sources (ASTM checklist 1988 Vol. 06.04: p.1209)

Inert or low-hiding pigments (fillers, extenders) Aluminium silicate (anhydrous), aluminium silicate (hydrous), barium sulphate, calcium carbonate, diatomaceous silica, magnesium silicate (talc), pumice, wet-ground mica.

White pigments Calcium borosilicate, titanium dioxide, tri-basic lead phosphosilicate (this product will be withdrawn from the market), white-base lead carbonate, zinc hydroxy phosphate, zinc oxide, zinc sulphide.

Black pigments Bone black, carbon black, iron oxide (synthetic black), lampblack (soot).

Bronze and miscellaneous metallic pigments Aluminium powder and paste, bronze powder (gold), copper powder (anti-fouling), cuprous oxide (anti-fouling), mercuric oxide (anti-fouling), zinc dust.

Blue pigments Copper phthalocyanine blue, iron blue, ultramarine blue.

Green pigments Chrome green (pure), chrome oxide green, Phthalocyanine green.

Yellow, orange and brown pigments Strontium chromate, zinc chromate, chrome yellow and chrome orange, iron oxide (hydrated yellow), iron oxide (natural red and brown), iron oxide (synthetic brown) molybdate orange, ochre, sienna (raw and burnt), silichromate (basic lead), umber (raw and burnt).

Red pigments   Iron oxides (natural red and brown), iron oxide (synthetic brown), 'para' (pure red toner), red lead, toluidine (red toner).

Attributes of colour

Colour today is defined chiefly by using the *Munsell system* which evaluates colour in terms of:

Hue   The actual colour attribute, e.g. 'green' or 'red'.

Value   The brightness measured in terms of the degree of reflectance.

Chroma   The intensity of hue related to greyness.

BS 5252 provides a framework for colour reference which locates 237 colours using the visual attributes of hue, greyness, weight and lightness.

Hue   There are 12 horizontal colour rows of different hues with one row of neutral colours, as follows:

    02 red-purple
    04 red
    06 yellow-red
    08 yellow-red
    10 yellow
    12 green-yellow
    14 green
    16 blue-green
    18 blue
    20 purple-blue
    22 violet
    24 purple
    00 neutral

The two yellow-red rows allow for closer matching of browns with red pigments (06) and yellow pigments (08)

Greyness   *Greyness* allows for the grey content in colours, and colours are put into five groups from A to E as follows:

    A grey
    B nearly grey
    C grey/clear
    D nearly clear
    E clear

Weight and lightness   *Weight* is a subjective term for comparing colours and *lightness* is the attribute given to surface reflectance.

It is worth mentioning that the BS 4800:1989 range of colours is extremely limited as a list of 100 colours selected from the 237 colours established in BS 5252, and that major paint manufacturers produce a far wider range of colours. The Royal Horticultural Society have just re-published their own colour referencing system which was originally designed to assist in the accurate recording of colours of flowers. It was used widely by printers and traditional decorators for colour matching. As the colour of the samples is matched to the 'smooth and soft' textures found in nature the quality (intensity) of colour is very high.

Other sources of colour

Colour can also be created by the physical splitting of white daylight through a diffraction grating. Further cancellation of colours can be achieved by the interference of different wavelengths of light. There are many examples in nature. For example, the scales on a butterfly wing have fine grilles (so do the wing cases on beetles), and the resulting colour has nothing to do with pigmentation. The bright surface colours on insects and birds may be due to layers of moisture held between laminations or feather hairs. These layers can act as a prism: once a bird dies the moisture evaporates from the feathers, which lose their colour. These physical mechanisms to create colour can be copied in architectural coatings. For example the oxide film on some metals may be controlled so that the depth of the oxidized layer gives the effect of colour by interference. These colours are quite permanent as they are chemically stable in the environment.

Historical background

Paints were traditionally described in terms of their ingredients with advice and recipes for making them. Until the First World War it was quite common for contractors to have their own workshops in which the paint was mixed. The mechanics of hardening and adhesion were not properly understood so the emphasis was on recipes based on tried and tested coatings.

The early terminology used is still understandable today although the individual ingredients used may have changed. Rivington's *Notes on Building Construction* (1901) reads:

The paints used by the engineer and builder as a rule consist of a *base* (generally a metallic oxide) mixed with some liquid substance known as the *vehicle*; upon this, permanency of the paint depends.

In most cases a *drier* is added to cause the *vehicle* to dry more quickly, and a solvent is sometimes required to make it work more freely.

When the final colour required differs from that of the based used, the desired tint is obtained by adding a stainer or colouring pigment.

Bases white lead, red lead, zinc white, oxide of iron
Vehicles water, oils, spirits of turpentine
Solvents spirits of turpentine ...

The book then describes how to make all of these ingredients, recognizing to some extent their toxicity, e.g. on white lead paint:

> Where it is exposed, however, to the fumes of sulphur acids, such as are evolved from decaying animal matter in laboratories, and in some manufacturing towns, it soon becomes darkened by the formation of black sulphide of lead. It has also the disadvantage of producing numbness and painters' colic in those who use it.

It may sometimes be necessary to match traditional paints and, even though the information may be out of print, it is still accessible due to the meticulous recording of recipes at the time. However, the heavy reliance on lead ingredients, and some of the acids needed to modify them, should now be minimized or banned altogether. One often finds in conversion work when stripping paint back to the original layers the original lead-based coatings which are still a health hazard. Safety measures when undertaking this kind of work are essential.

Coating systems

Modern coating systems have the same basic constituents of:

Binders
Solvents or thinners
Pigments

Different types of paint have different constituents, according to their use. *Initial coatings* or *primers* should have good *adhesion* and be able to immediately protect the material against natural decay. Metal

Table 3.2 Minimum thicknesses of a single coat

Situation	Thickness (μm)
Clean country air	125
Polluted city air	180
Corrosive sea air	250
Industrial air	300

Table 3.3 Typical coating thickness

Paint type	Thickness (μm)
Acrylic resin	25–30
Alkyd resin	25–30
Chlorinated rubber	150
Polycyclic latex	15–25
Traditional oils	35–40

primers will often have rust inhibitors. Some primers carry pigments that help achieve this, for example red lead or zinc powder. The primer should also cover imperfections and bridge minute cavities that could initiate corrosion fatigue, *Top coats* are chosen not just for their brilliance of finish but for their hardness against erosion, and stability against aggressive environmental conditions.

Coating thickness   Initial protection is given by the thickness of the coatings and is dependent on the severity of atmospheric conditions (Table 3.2).

Minimum coating thicknesses   As film thicknesses individually are so thin this means that layers of coats must be built up for adequate protection (Table 3.3).

Primers must adhere well to a substrate and provide an efficient base for adhesion by undercoats. Their importance cannot be overstressed, and failure of the primer will immediately mean failure of the whole coating system. Unfortunately primers vary in their performance and must be specified adequately. They also have to take substantial strains, especially in timber where seasonal movement can give a strain figure of up to 10% for the coating. 'Permanence' can vary in different primers and failures have been recorded by the Building Research Establishment (BRE) between one week and six months later. General points to note are that slow drying primers give better results, primers even without lead content can still be perfectly adequate and any timber should be treated with preservative prior to priming by immersion or vacuum methods. Permanence can be a particular problem with factory-primed joinery and care should be taken to ensure that suppliers are adhering to standards specified.

3.2 Organic coatings

These are the modern *carbon-based* polymer coatings.

Organic coats and their classification

Paints are generally classified according to how they harden, transforming from liquid (solvent and pigment) to solid (matrix composite) either by evaporation of solvent or by a chemical reaction.

Evaporative drying   The film hardens with no chemical reaction taking place. The paint constituents are suspended in a *medium* commonly referred to as the *vehicle* and evaporation occurs of a particular solvent with no chemical reaction.

Polymer constituents commonly found in a solution of organic solvents include:

Cellulose group   Cellulose nitrate; cellulose acetate; ethyl cellulose.
PVC   Polyvinylchloride   and   polyvinylchloride-polyvinylacetate co-polymer.

These are commonly embraced in the term *vinyl resins*. PVC is insoluble in ordinary solvents and highly resistant to alkalis and acids as well as oxidizing agents. PVC can be degraded by heat and light. It is a linear thermoplastic and commonly used as a co-polymer with vinyl acetate. It has a tendency to crystallinity.

Acrylic   Poly(methylmethacrylate)   Acrylic resins have a high softening point (T_m) and can be easily converted to thermosetting resins by the addition of other monomers, e.g. styrene. It is soluble in a variety of solvents and is a common binder in paints.
Polyvinyl butryl   Polyvinyl butryl is used in metal pre-treatment primers (etch primers) and helps give good adhesion for undercoats.
Rubber and chlorinated rubber

Note: The loss of solvents is a health hazard. Instructions should be followed and work carried out in factory-controlled conditions or well-ventilated spaces. Failure to follow these instructions can lead to permanent damage to living organisms and tissue.

Where the polymer constituents are dispersed in water the paints are either *emulsion* types with one solution suspended in another, or *latex* types with fine particles dispersed in water. As the water evaporates particles come closer together and coalesce to form a continuous film. Plasticizers are often added and work by depressing the T_g (glass transition temperature) of the main polymeric compounds so that the film stays in a rubbery state at ambient temperatures. Paint films lose their plasticizers over time as the plasticizers migrate into the atmosphere. This causes the film to become hard and ultimately brittle. The setting of these films can be critically dependant on temperature. Hardening at low temperature (i.e. close to the glass

transition temperature) could cause hardening of dispersed particles before they coalesce to form a film. This would give a 'powdery' finish, instead of a continuous hardened film which results from the even solidification of an even liquid coating. The most commonly used polymers are:

Poly(styrene-co-butadiene)
Poly(vinyl acetate) and co-polymers
Poly(methylmethacrylate) and co-polymers
PVC organic solvents

Convertible coatings   All of these coatings undergo a chemical reaction when they change their state from liquid to solid.

Oxidative polymerization   In this process, the films actively combine with oxygen molecules in the atmosphere during polymerization. This includes most of our traditional oil paints and give chiefly crosslinked polymers. Examples are:

Fatty drying oils
Oleoresinous coatings
Air-drying alkyds

Oleoresinous binders are vegetable in origin (e.g. linseed oil from the flax plant and oils from the olive, soya, castor, coconut and cotton plants). By the adding of natural or synthetic resins and omitting pigmentation, these paints will give oleoresinous varnishes. To speed up the hardening process, salts of lead, cobalt and manganese may be added, but any of these accelerators can also cause overhardening and give a more brittle film.

The alkyds (polyesters) contain between 35% and 80% oleoresinous components. They can be degraded by alkalis. Different categories of alkyd resins vary according to the ratio of oleoresinous components used. Short-oil ratio alkyds are used in mixtures which can be cured at high temperatures, i.e. mixtures suitable for baking or stoving processes. Those oils with medium- and long-oil ratios can be cured at ordinary temperatures. Categories known as *modified-alkyd resins* can incorporate chlorinated rubber to improve corrosion and fire resistance, and to accelerate hardening; cellulose nitrate to increase hardness; amino resins to increase hardness and improve resistance to alkalis; and lastly silicone resins to improve water and heat resistance.

These coatings harden by a chemical reaction within the film. Conversion (or curing) is often assisted by the input of external energy, i.e. ultra-violet light. Examples are:

Epoxy EP
Unsaturated polyester
Polyurethane
Urea and melamine formaldehyde

Water-based paints

As exposure to VOC's is now recognized as a health hazard, very high air changes are required of 7–10 air changes per hour (ach) to ensure that recommended exposure limits are not exceeded. Water-based paints have lower emissions and should be specified for all interior decoration. Paints that use solvents should be restricted to external use where there is some superiority in performance.

The BRE has carried out trials of water-based paints. The results show that these compare favourably with organic-based paints, with a life of 5–8 years which can sometimes extend to over 10 years. See BRE information paper No IP 4/94 which gives application and properties. BS 5082:1993 is the standard for water-based wood primers of comparable performance to the old lead-containing primers. The only disadvantage with water-based coatings is their application in poor weather conditions: rain can damage the wet film and frost can affect it during the drying process, to the extent that it has to be stripped off and recoated. Paint can also harden incorrectly if exposed to hot conditions or intense sunlight, resulting in a poor quality coat. Some high-permeability coatings have affected the durability of cladding timbers on buildings in Scandinavia. Water-based treatments do not adhere to fresh putty. Putty should be left to skin-harden or be sealed with organic paint. There is a lot of experience of using water-based paints in America, where VOC content in paints is limited in some states in order to improve air quality. ASTM 3129:1992 deals with the relevant tests for Exterior Latex House paints and ASTM D3960:1993 with determining the VOC content of paint.

The Hazard Information and Packing Regulations 1993 require that all hazardous constituents in coatings are identified and that proper protection is used in the application of these products. However, even water-based paints may contain many organic compounds to which there are occupational exposure limits. These compounds include:

Ethylene glycol
Propylene glycol
Methoxy propoxy propanol
Phenoxy ethanol
Naphthalene

Aromatic hydrocarbons C10/C11
White spirit

All of these compounds improve the flow of the paint and help the coalescence of particles.

It is difficult to formulate water-based paints that match the properties of organic paints. Manufacturers in Denmark, France and Germany have produced new water-based products and these are being imitated by manufacturers in the UK. There is renewed interest in early paint systems that used natural oils, such as linen seed oil, safflower oil, wood oil and castor oil, as well as resins from larchwood and pine. It is possible to formulate thinners from pure plant oils from citrus peel and pine for cleaning brushes as well as dilution.

In England, Environmental Paints Ltd have developed products which also use natural pigments. Inevitably colour ranges are more limited.

Clear finishes: Natural finishes or varnishes

Varnishes often consist of the same ingredients as a paint film without pigmentation. Although they provide a high gloss finish to a material and some protection against solvents especially water, they often cannot filter out the ultra-violet radiation components in daylight that a pigment would normally absorb. This leads to degradation of the film by excessive crosslinking, making the film more rigid and hence brittle. The film ceases to be able to move at the same rate as the substrate, and once crazing begins, light waves are scattered along the edges of cracks causing opacity.

Examples of clear finishes in the non-convertible category of films are shellac, bitumen and cellulose nitrate (commonly known as *cellulose lacquers*). They are non-convertible as they can be re-dissolved in their original solvents. This is an advantage in the repair of French polish which is simply shellac dissolved in methylated spirits. Convertible polymeric coatings are more stable but once damaged have to be replaced in their entirety.

Stains

Stains can be a one- or two-coat application. If made up of two coats, it is likely the first coat will have a lower solid content for greater penetration.

Low solids <30%
High solids >40–50%

Stains are more economical than paint films as fewer

coats are applied. The higher the content of solids the greater the ability of the film to resist degradation by ultraviolet light as there is a greater opportunity for the absorption of light.

The stain usually consists of a resin vehicle, pigmentation and a fungicide. Very dark colours, e.g. black, brown, dark blue, may act as a heat sink and cause excessive shrinkage in the timber below. The mid-range of colours, e.g. blues or greens, can give very effective protection without causing complications from heat gain.

A major disadvantage of stains is that their permeability is often high relative to paints. This allows the passage of water vapour but can increase the movement of jointed components. Preservatives (e.g. pentachlorphenol, tributyl tin oxide) incorporated into stains are not effective as long-term protection to timber, as they only prevent mould growth on the finished film and should really be regarded as anti-fungicidal. Separate impregnation of timber approved British Standards prior to staining should always be carried out.

Stained treatments still require good maintenance and this is carried out usually at the first signs of loss of gloss or sheen and before any noticeable deterioration. Once colour deterioration is seen damage will already have occurred to the timber substrate and there will be loss of adhesion of subsequent coatings of stain. This can only be rectified by substantial preparation of the timber.

The aim in staining is for a minimum life of five years and some stains give a performance (often with two coats) that can be twice as long. Currently available stains do not give deep penetration of the substrate and the penetration is no deeper than regular paint systems.

Note: In external joinery, windows glazed using putty rely on paint for protection. When staining timber joinery a butyl, polyisobutylene, polysulphide or similar adhesive should be used instead of putty. These compounds react unfavourably with compatible paint films used internally and in some situations neoprene gaskets may need to be specified. This is common practice for proprietary window products in Scandinavia.

3.3 Inorganic coatings

Inorganic coatings are not carbon-based and their basic constituents are inorganic including metallic ones. They are often used as priming coatings used to protect metals.

Conversion coatings

Conversion coatings undergo a chemical reaction when applied and can physically combine with a metal substrate. This makes them ideal as primers. They often form an oxide coating: a type of controlled corrosion. Another type forms a new metallic compound at the surface of the metal by causing the metal to disassociate into positive ions which combine with the metallic element in the coating. The pH usually rises as the new compound is formed so changing the immediate environment around the metal from an acidic one (which encorages corrosion) to a more passive alkaline one. The coatings formed are very thin, perhaps only 2 microns thick. It is important that further coatings are applied quickly.

Integrated controlled corrosion

Some metals (especially steel) contain additional alloying elements which ensure that surface corrosion is limited and a corroded layer builds up which protects the metal from further corrosion. All stainless steels use this technique: they contain a small quantity of chromium, which oxidizes rapidly in the atmosphere to form a strong coating that can be polished. *Cor-ten* contains approximately 0.4% copper and will form a weathering steel. (Care should be taken in the use of cor-ten, as rainwater run-off can stain adjacent materials). *Phosphorus* is also commonly added to alloys to bring about controlled corrosion.

The compounds of coatings which cause controlled corrosion must be comparable with the parent metal in terms of their molecular size and atomic spacing in order to ensure a good degree of inter-metallic bonding. Incompatibility in size will cause stress-cracking, and the new exposed surface will cause further corrosive deposits to grow. This can give a patchy and unsightly layering to the material. It is thought that in the additional alloys some of these finishes cause 'growth poisoning' and inhibit the normal formation of a crystalline oxide layer of the parent metal.

The oxidation of iron (rusting) is unacceptable because the corrosive deposits formed are large compared with the parent metal and contain voids, which further expose the metal and lead to continuing and progressive corrosion through the whole body of a section.

Enamels

Vitreous enamel coatings applied to steel are usually extremely thick, up to 0.8 mm. They are applied as a powder, which is fired when it fuses to the metal.

Although they are resistant to most chemicals, especially acids, they are susceptible to alkali attack. (Note the corrosion of bath enamel in hardwater areas.)

Cement-based paints

Some paints contain white Portland cement with or without pigments, water repellants, fillers and plasticizers. They are water-based and can only be used on brickwork, concrete or renders. They are being replaced now by proprietary masonry paints which are acrylic emulsions with alkali-resistant large-scale pigments and fungicides.

3.4 Special coatings

This is a new category of coatings including new developments in ceramics as well as some traditional ceramic coatings.

Research and development

Much research and development money has been spent on engineering organic polymer coating systems, and manufacturers are keen to maintain their use within the building industry. Technology may improve but the industry is built on a premise of continual renewal and constant demand; specifications of these finishes are to some extent self-perpetuating unless other options are considered.

Vulnerability

As paint coatings are only microns thick they are vulnerable unless well-engineered for stability under stress; whether through dimensional movement or temperature change from $-10\,°C$ to $+40\,°C$. Paints must also resist direct abrasion and ingress from water molecules.

Adhesion

The film should not peel away from its background. In a previous section (2.3) we examined adhesion and should reconsider the following areas:

    Surface wetting
    Thermodynamics
    Surface chemistry and possible bonding

Degradation

Degradation is a major problem with organic coatings. Most research and development is devoted to the stabilization of unstable materials. For instance, ultra-violet light causes reactions in polymers and extensive cross-linking that can cause embrittlement.

UV light can also break polymers. UV stabilisers are added to combat this by changing their structure when absorbing energy, and changing back when the source of UV is removed, emitting heat at the same time. Water absorption by plastics causes swelling as the films take up the water or moisture vapour. The material stiffens and cannot flex as easily under stress. It becomes vulnerable to hairline cracking, sometimes referred to as embrittlement by stiffness.

The faults just listed are known to occur with all organic polymer coatings. Instead of investing time and energy in making and using products from materials that are inherently unstable in our atmosphere, other options should be investigated. In a particularly aggressive environment or one where we have a very high expectation of a surface coating finish, we should specify instead:

- Not using finishes which are fraught with problems and lead to failure of the building fabric (timber, etc.) if the finishes are not well maintained.
- The use of finishes that are already environmentally stable, i.e. relatively inert and not prone to decay or deteriorate.
- A finish that is able to form a chemical bond with the substrate.

Priorities

The problem is most urgent for metals. Their rate of decay is not only costly in terms of renewal but also dangerous. The failure of films will lead to the initial corrosion of steel through intergranular corrosion, possibly leading to more severe failure of the material, most certainly significant in highly engineered structures.

The most stable coatings currently known are ceramics. They are inert and will not change in the atmosphere as they are already in an oxidized state. There is also a proven compatibility between metals and ceramics: most ceramics have metallic elements as part of their composition. The following categories of coatings show the variety of possible uses and applications of ceramics, some of which are still at the development stage.

Glass type coatings Good surface 'wetting' of the metal to be coated is important. So is cleanliness of the metal surface. Although the metal may be cleaned to a state of brightness with the complete removal of oxidized material, the metal may then be left for a

short while for a controlled depth of oxidation to take place. This actually assists adhesion. Sometimes intermediate oxide coatings are applied to bright surfaces which may require an intermediate firing. After surface preparation, powdered glass is applied to the surface and then heated. It is important that the glass types used can liquify at temperatures that do not cause phase transformations in the metal substrate. It is important that the glass and metal do have similar coefficients of thermal expansion. The application of such coatings requires close control. The manufacturing process should be able to deal with large components e.g. cladding panels. Signs and steel and cast-iron baths are commonly enamel-baked. The technology of these coatings is improving as the controlled oxidation of the parent metal is optimized, and the thickness of the coating is reduced. Old porcelain ceramics were very vulnerable to chipping because of the thickness of the coating.

Unfortunately the surface of ceramic coatings can deteriorate, which leads to loss of hardness and the surface imperfections can result in a significant loss of strength. Some common environmental conditions degrade ceramics. Hard water (which is alkaline) can permanently alter the structure of Si-O configurations in the ceramic matrix and dissolve it. The problem increases with the concentration of OH^- ions. Although resistant to most acid solutions, ceramics can suffer severe degradation from hydrofluoric acid. As some detergents also generate alkaline species they should be avoided. Silica deposits may also be precipitated.

Composition coatings inhibiting migration of iron   These coatings were originally developed and patented for the food industry: the taste of canned beer is altered by a concentration as low as 60ppm of iron. The problem is the sealant around the ends of the cans: Beer is acidic and if it comes into contact with the scored ends of the can, iron atoms from the can dissolve into the beer. The inner coating used now is an organic enamel based on polyvinyl chloride, with a plasticizer. A metal oxide (such as aluminium, zinc, magnesium or calcium) is now added to this enamel coating and it inhibits soluble iron migration by acting as an acid scavenger.

Weather-resistant coatings   Research is currently being carried out which aims to add components to paint coatings that will make effective bonds with metals and provide an environmentally stable coating. Calcium and strontium oxalates, and group 1B or IIB metal oxides or sulphates mixed with water soluble silicates look promising. The greatest difficulty is in finding compounds that will bond effectively at ambient temperatures. Most work undertaken so far has concentrated on the fusion of ceramic and metal interfaces by heat, a laser or an electric field. In effect, this is a similar process to the making of particle composites. This is an expensive procedure and not a site application.

The search for effective compounds has now come back to some materials that were used traditionally and referred to as pigments. These pigments were often inorganic compounds suspended in organic matrices or solvents, but their properties were never properly evaluated except as colouring agents. The best known reactive compounds made very effective primers, such as red lead.

Long-life inorganic coatings   The major drawback of organic coatings is their limited life. They have a limited 'pot life' and as films they degrade over a long period of time. Development work is being carried out on coatings with inorganic composites which have a long pot life. They contain, for example, condensed phosphates in an aqueous silicate solution.

Coatings resisting sagging   Organic coatings must stay in a 'rubbery' phase and be light in weight, so that sagging will not occur before proper hardening has taken place. When thick coats need to be built up for heavy-duty protection, special coats using water-soluble silicides, or silica solvents, with a hardener solution containing phosphate and aluminium are now possible alternatives.

Textured finishes

Hammer finishes   Hammer films are popular today because of their durability and physical appearance. They provide a 'beaten' dimpled surface (Fig. 3.4). The whole effect relies on the method of evaporation of the solvents used in the paint. This happens at great speed causing a vortex effect and Bernard cell forma-tion. Bernard cells are charcterized by hexagonal geometries which take least amount of surface energy to form. These films contain metallic powders, usually aluminium, and the flake-like particles, when rotating in the vortex action, align themselves with the sides of cell walls and intensify light reflection. These finishes can be sprayed and should not be applied thickly (approximately 25 microns) as this can inhibit good cell formation and leave the lower part of the film less rigid. Hardening time is fast, in the region of 15 minutes.

The binders used include styrene-containing alkyds,

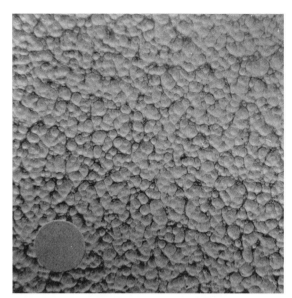

Figure 3.4 Hammer finish on the surface of a gas fire appliance.

vinyl toluene co-polymers, chlorinated rubber, polyurethanes, epoxide esters and nitrocellulose.

Wrinkle finishes These kinds of finishes were commonly used on metals after the Second World War for small artifacts, such as typewriters. They are very durable and should be applied by spraying. During hardening a smooth crosslinked film forms on the surface, the lower layers are starved of oxygen and individual monomers are absorbed by the surface skin which swells. As the skin is floating on a liquid surface and still attached to it, the swelling deforms into a wrinkle formation. A similar effect can be inadvertantly achieved when using a paint that softens an underlying coat: the new top coat wrinkles because of the same mechanism and is of course a defect.

Binders include oleoresinous varnishes, modified phenolic resins and alkyds, and vinyl resins. An alkyd wrinkle is approximately 50 to 80 microns in thickness which is significantly larger than individual coatings in paints (of the order of 25 microns) and so provides more effective protection.

Intumescent paints These paints are used in particular to provide fire-resistance. On heating, components in the film react giving off gaseous products which cannot escape due to the strength of the polymeric materials. The gases evolved are non-flammable, such as carbon dioxide or ammonia. The coating then often expands by an order of a hundred fold converting from a film to a thick rigid foam. Some

paint films can be as thick as 1.5 mm (1500 microns). The thickness will depend on the degree of fire protection required and the structural member to which it will be applied. For example, columns are required to have a more substantial coating than tie struts.

Note that these coatings are often water-sensitive and may not be suitable for external use. Individual requirements should be checked with the manufacturers concerned.

Powder coatings There are two main categories of powder coatings: *thermosets* (epoxy resins) and *thermoplastics* (chiefly acrylics and polyesters).

Powder coatings are often a mixture of polymeric materials, commonly referred to as 'co-polymers' which eventually form hard resins. They are ground into fine particles and applied to metals (often electrostatically for an even coating) and then heated in ovens, which causes the particles to coalesce. These polymers are commonly thermosets and show good powers of adhesion. Thicknesses between 80 to 90 microns can be achieved and they set quickly, commonly between 6 and 30 minutes. They were originally developed as an alternative to cellulose paints which were in widespread use in the automobile industry. Cellulose paints give off toxic solvents (VOC's) and, as we have seen, their use is being controlled by legislation. In addition most organic solvents are derived from crude oil, which is steadily increasing in price encouraging the development of alternative products.

Thermoplastics give a characteristically smooth and glossy film surface but are not as durable and do not bond as well to metal substrates or show as great a resistance to organic solvents as thermosetting powder coatings. However, thermosets have a poorer surface finish; due to setting during heating before attaining complete fluidity, they have a slightly rougher surface and hence less gloss.

Both categories of powder coatings have a complex mixture of ingredients which include the following:

Flow-control agents
Anti-static agents
Pigments
Plasticizers

In fact these coatings are better classified as products of particle composite technology (See Section 4.0).

These coatings are sometimes similar to conventional paint systems with discrete layers. They are known as *multi-layer powder coatings* and can still be applied in a one-spray operation which contains opposing electrically charged powdered components.

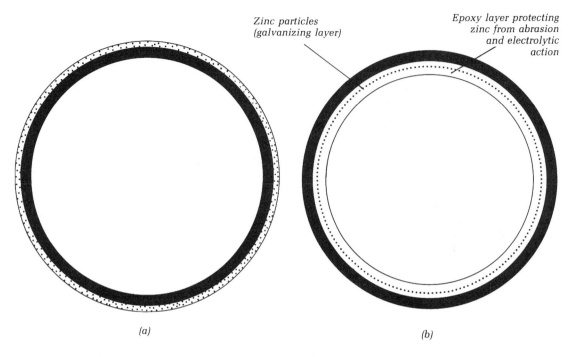

Zinc particles (galvanizing layer)

Epoxy layer protecting zinc from abrasion and electrolytic action

(a)   *(b)*

Figure 3.5 Multi-layer coatings on the interior of pipes. Pipe exterior (a) – the powder mixture is blown on to a pre-heated surface which is above the T_m of the powder (70–120 °C). There is instant fusion and good resistance to future de-bonding. The epoxy here is polyglycidal ether with barium sulphate and calcium carbonate filler in an epoxy resin matrix. Pipe interior (b) – this coating is applied in one spray treatment. Different powders have opposing electrical charges which produce a separation into layering. This means that the coating will have appropriate properties for the substrate as well as the exposed surface.

These have clearly separate functions. For example, in pipe interiors zinc particles are applied first to give a galvanizing layer to the steel. A surface layer of epoxy resin is then applied to protect the zinc coating from abrasion and electrolytic corrosion (Fig. 3.5). Sometimes components are sprayed in such a way that particles become suspended in the matrix, commonly an epoxy resin.

There is an alternative to spraying powders onto cold substrates and then heating them. If the material, usually metal, is heated first, then, as spraying occurs, fusion is instantaneous (70–120 °C). This gives better resistance to de-bonding in the long term.

Anti-fouling paints Anti-fouling pains must be used on marine structures. They release toxic compounds as part of the anti-fouling mechanism. These products must be handled with care and not used in situations where maintenance of the ecological balance is a major consideration.

Fungicidal paints The prevention of mould growth can be a factor in the choice of paint films. This applies not just in housing but in situations generally where the microclimate favours the growth of micro-organisms. Such paints such not be used as a remedy for existing problems with ventilation and condensation. The Building Research Establishment publish lists of coatings that have proved effective in preventing mould growth for three months. It is unlikely that mould growth will begin again. There are also anti-condensation paints that prevent mould growth. These probably work on the principle of manipulating the surface tension of water droplets, making it impossible for spheroids to form and so keeping the water in a vapour state. See BRE *News of Construction Research*, October 1986.

3.5 Preparation of substrates

Prior to painting, all surfaces should undergo treatment to eliminate dust, grease, and old defective paint. Backgrounds should be thoroughly cleaned, not simply brushed down, as otherwise the paint will only be adhering to loose particles and not to a solid substrate. Preparation is the most important part of the coating process. New plaster and render must be left to fully hydrate: there may still be some additional moisture loss. Checks should be made on the moisture content of the surface to be coated:

1. by hand-held instrument
2. by sealing off an area of approximately 300 mm square (recommended in BS 6150:1991) and taking measurements in the centre with moisture sensitive paper
3. using a hygrometer (as described in BRE Digest No. 55). This method can be applied to all large-scale surface finishes that involve wet trades. ASTM D4263:1988[3] provides an equivalent moisture test which uses 450 mm square sampling surfaces.

Removal of old material

Burning off   Existing paint films can be removed by heat treatment. Blowlamps can produce burn marks if used carelessly, but hot-air paint film softeners are less likely to mark timber.

Solvent and chemical strippers   These solutions are very aggressive chemicals and should not be used in confined spaces. Instructions should always be carefully followed. If these solutions are used the surfaces should always be washed clean. See BS 3761:1986 *Specification for solvent-based paint remover*.

Physical removal of paint films   Sanding timber, working from coarse to fine grade, produces good results as loose fibres and debris are removed from the timber substrate. Sandblasting is also effective, although samples should be tried first. The coarser the particles used, the higher the rise of grain as softer pulp is removed.

Removal of fungal and algae growth on paintwork and related coatings   Exact guidance for this can be taken by ASTM standard D4610:1986 from which the following is taken or from BS 6150:1991.

This is a problem especially in micro-environments which foster biological growth. Sodium hypochlorite ($NaClO_3$) solution can be used as 5% aqueous solution in conjunction with tri-sodium phosphate or alone.

The surface is tested first with a drop of the sodium hypochlorate solution. If it bleaches, micro-organisms are present. The surface should first be washed with the tri-sodium phosphate solution, thoroughly washed with clean water and then one part sodium hypochlorate solution should be diluted with 3 parts water and applied for 10–15 minutes before thoroughly rinsing the surface again with clean water. After the surface is completely dry it can be painted.

Timber

Most timber used in building has to undergo some form of surface protection.

One of the chief problems with timber is that it absorbs moisture from its surroundings and this creates an environment conducive to life. If the timber is dead, insects or fungi will start to break down the structure of the timber and return it to the natural food cycle. This is an inevitable process and our use of timber simply delays the cycle for a time, depending on how impenetrable our coating systems are, or how lethal the constituents are that they contain.

The second major problem with timber is that all untreated timber will bleach and eventually turn grey under the effect of ultra-violet light. Very few timber species stay stable in these conditions: the ultraviolet radiation destroys the lignin content in the timber. When this binding material, which is also a polymer, degrades only loose cellulose fibres are left and the surface de-laminates.

For general guidance see BRE Digest No. 354:1990 which gives a summary of the main reasons for timber deterioration and how paint films and preservation techniques delay decay.

Preservation   Before any surface coatings are applied it is essential that the timber has already been treated against decay, and there should be evidence of certification by the timber supplier, contractor or joinery subcontractor. The most effective treatments contain salts that are toxic to fungi and pests. Double-vacuum methods of absorption of the liquid are the most effective. See BS 4072 Parts 1 and 2 for treatments for external woodwood in buildings (excepting ground contact) which use copper chrome arsenate (CCA) water-borne preservatives, applied using the vacuum method; and see BS 5707:1986 for pentachlorophenol organic solvents. Timber with ground contact should be treated with creosote preparations as detailed in BS 5589. This standard also lists durable timber species where the use of preservatives may be avoided.

If preservatives are used, checks should also be made for compatibility and should cover:

- The paint/coating system and the preservative in use.
- Any effect on fixings e.g. thin-gauge steel gang nail-plates, ferrous nails and screws

See BS 5589:1989

It is advisable to specify all metal timber fixings to be corrosion-resistant, as systems of preservation can create a mildly acidic environment. This will act as a weak electrolyte, increasing the risk of eletro-chemical corrosion of ferrous fixings.

Moisture content   The moisture content should be checked prior to coating. This is relatively easy for a contractor to do using a hand-held instrument. It should not exceed approximately 18% for general woodwork, but in centrally heated environments the moisture content may go down to as low as 10% depending on the temperature to be maintained in the building all year round. (See *MBS: Materials* Chapter 2, Timber).

Generally the procedure for painting timber should follow section 25.4 in BS 6150:1982. The following is a summary of the work.

General repair   The wood should be checked for soundness with a knife or hat pin. A strip of Sellotape may be applied and pulled away to establish if there are any loose flakes (Fig. 3.6). If this is the case the wood must be sanded or planed to achieve a smooth sand surface. If the wood has been repaired and pieced in, check that no surfaces are proud or need further sanding. Sandblasting is also an option but will sometimes remove the weaker and pulpy cellulose tubes giving the wood a ridged character.

Figure 3.6   Using Sellotape to detect loose and fibrous material. Although this is an extreme example, it is a very useful site test for the surface preparation of timber. Any loose flakes picked up indicate a surface too poor to accept coatings without further preparation.

Sanding   All surfaces should be sanded with fine glass paper in the direction of the grain. (Coarse grades will only detach sound fibres and damage the surface.) Surfaces used to be dusted off, but today hoovering is more convenient and will remove dust from the painting area. At this stage some oily hard woods (teak, aformosia, gurjun and makore) can be washed with white spirit, which helps to prevent loss of adhesion on these particular timbers.

Knot treatment   *Knotting*, the stabilization of knots, must be carried out to seal the consistent seeping of natural resins from the end-grain features of branches or twigs, etc. Preservative will also seep from these end-grain features and knots have to be sealed to prevent any damage to new coatings. See BS 1336: 1988 for the specification of 'knotting' (applying shellac in methylated spirits). Aluminium primers may be adequate to seal knots indoors but not externally, where under intense sunshine resinous knots may destabilize.

Priming   It is essential that primers are applied soon after the timber has been dried to its correct moisture content. End-grain must also be fully primed as the ingress of moisture along open cellulose tubes can be as much as a hundred times that from the side of material. Most paint failures are due to the poor quality of the timber surface. There is no point in painting over rotten and decayed wood: it should be systematically prepared and if necessary stripped out and replaced.

BS 6150 outlines three primer types for use in wood work (Table 3.4). The first is for a low lead (less than 1%) oil-based primer to BS 5358:1993 *Specification for solvent-borne priming paints* for woodwork and for general use. The second priming system is for an aluminium wood primer to BS 4756:1991 *Specification for ready-mixed aluminium priming paints for woodwork* and is suitable for resinous woods. The third is for a water-thinned primer to BS 5082:1993 *Specification for water-borne priming paints for woodwork* and has rapid-drying qualities. All primers must be compatible with the system of finishing coats to be used, and a check should be made on the primer manufacturers' compliance to a relevant British Standard. Finishing coats should be applied as soon as possible after priming and after checking the moisture content of the substrate.

Stopping and filling   Cracks and holes should be filled after priming with oil-based compounds which are more compatible with finishing coats. Putty is a traditional filler and should conform to BS 544:1969.

Table 3.4 Primers for wood

Description	General composition	Characteristics and usage
Solvent-borne primers to BS 5358 [A]	Drying-oil/resin type binder. Non-lead pigmentation	Primers of this type are equivalent in performance to the traditional 'pink' (white lead/red lead) wood primers. They are suitable for general use on wood, not highly resinous and not treated with copper naphthenate preservatives or creosote; also for fibre boards and wood chipboards, not fire-retardant treated. *Colour*: Typically white or pink
Ready-mixed aluminium primers to BS 4756 [B]	Drying-oil/resin type binder with aluminium pigment	Alternative to **A**, and more suitable for woods which are resinous or have been treated with copper naphthenate wood preservatives or creosote. May also be used as primers for fibre boards and wood chipboards (not fire-retardant treated) and as 'primers' for surfaces that have been coated with bituminous materials. *Colour*: Aluminium
Water-borne primers to BS 5082 [C]	Emulsion-type binder (typically based on acrylic polymer). Non-lead pigmentation	These primers dry more rapidly than types **A** and **B**, usually allowing same-day recoating under good drying conditions. They are not flammable and permit tools and equipment to be cleaned with water. Their durability without top coats on exterior exposure is equivalent to types **A** and **B** but, as they are more permeable, they may be less effective in excluding moisture from primed joinery stacked in the open. They are more prone to raise grain than oil-based primers. They may also be used as primers for fibre building boards and woodchip boards (not fire-retardant treated). *Colour*: Typically white, light grey or pink
Semi-transparent primers (basecoats) [D]	Solvent or water-borne	Low opacity primers for use under conventional paints or exterior wood stains. They are of low durability and should be overcoated with minimum delay
Transparent preservative primers [E]	Unpigmented solution of resins and fungicides	First coat of many proprietary wood finish systems, generally unpigmented and containing fungicides. Should be overcoated without delay

Source: Based on Table 1 BS 6150:1991

If applied prior to priming its binder can be absorbed into the timber leaving a friable surface to paint over. Water-based fillers will need re-priming prior to painting. It may also be necessary to 'flat down' with fine glass paper prior to painting.

Undercoats   Two coats should be applied for external work, but this is optional internally and depends on the final quality. Undercoats should be flatted down if a high gloss is required on the top coat.

Top coats   Externally, two coats are ideal. If two are

applied, one undercoat may be omitted. Internally, one gloss coat is usually sufficient unless very high quality work is required. Gloss finishes are more suitable for use externally as they are more effective in allowing run-off of moisture.

Paint systems for wood, used internally or externally, are listed in Table 3.5.

Maintenance and cleaning   It used to be common practice to wash down paint surfaces with soapy water two years after application. They were then 'leathered off', in a similar way to cleaning windows. This work

Table 3.5 Paint systems for internal and external wood

Application	Requirement for preservative treatment*	Primers (see Table 3.4 for types)	Finish systems	Typical life to first maintenance†
Window joinery, softwood, internal and external	Essential	Solvent-borne [A]. Aluminium [B] (preferred for resinous woods, e.g. Douglas fir, and possibly for timber treated with naphthenate preservatives). Water-borne [C] (compatibility with water-repellent organic solvent preservatives should be checked). Semi-transparent basecoats) [D]	*Gloss external* one or two coats undercoat, one or two coats alkyd gloss finish, to give a three or four coat system as required	3–5 years
			Gloss internal one coat undercoat, one coat alkyd gloss finish	5 years or more
			Mid-sheen (semi-gloss) internal two coats alkyd mid-sheen (semi-gloss) finish	5 years or more
			Matt internal two coats alkyd matt finish	5 years or more
		Transparent preservative [E]	Multi-coat paints formulated for improved performance	4–7 years
External cills, hardwood	Optional but necessary if excessive sapwood present	Aluminium [B]. Filling necessary with open-grain timber	As for *gloss external* above	Generally as for window joinery but depending on species and nature of timber
Doors and frames, internal				
Softwood	Optional	As for window joinery	As for window joinery	As for window joinery
Plywood	None	As for window joinery. Filling recommended on open-grain veneers	As for window joinery	As for window joinery
Doors and frames, external				
Softwood	Desirable	As for window joinery	As for window joinery	As for window joinery
Plywood	Optional; required if it contains non-durable species	As for window joinery. Filling recommended on open-grained veneers or if checking has occurred	As for window joinery	As for window joinery
Skirtings, softwood	Optional	As for window joinery	As for window joinery	As for window joinery

continued

used to be undertaken by the decorator and it increased the service life of the paint films, especially externally. Sooty dust dissolved in rainwater creates an acidic solution which degrades paint.

Natural finishes for wood For timber the main natural finishes are varnishes, stains and French polish

(section 2.12). There are other methods (Table 3.6).

Oils These treatments enhance water-resistance but a good number of coats have to be rubbed in. They must be allowed to harden between coats.

Wax polishes

Rubbed undercoat finishes

Table 3.5 continued

Application	Requirement for preservative treatment*	Primers (see Table 3.4 for types)	Finish systems	Typical life to first maintenance†
Cladding, barge-boards, fascias and soffits				
Softwood	Required by Building Regulations for some species (not Western red cedar)	As for window joinery. Back-priming desirable	*Gloss finish* As for window joinery	As for window joinery
Plywood	Optional; required if it contains non-durable species	Solvent-borne [A]. Aluminium [B]. Paper overlay desirable and may allow primer to be omitted	*Textured coatings* Finish types may be suitable for use on timber cladding; consult manufacturers	Up to 10 years
Gates and fences‡				
Softwood	Essential	Solvent-borne [A]. Aluminium [B]	*Gloss finish* As for window joinery	Up to 5 years, depending on design and degree of exposure
Hardwood	Desirable and may be essential in some circumstances; see BS 5589	Solvent-borne [A]. Aluminium [B]	*Gloss finish* As for window joinery	Up to 5 years, depending on design and degree of exposure

Note: Reference should also be made to BS 6952: Part 1 for guidance concerning the classification and selection of exterior wood coating systems.
* Reference should also be made to BS 5589.
† Life expectancies shown assume application to dry, sound woodwork which, if necessary, has received appropriate preservative treatment and are based on performance in 'moderate' environments.
‡ Unless painting is necessary for appearance, consideration should be given to preservative treatment of gates and fences initially and for subsequent maintenance.

Source: Based on Table 11 BS 6150:1991

French polish
Acrylic emulsions
Liming   Liming, used traditionally to finish oak, has been enjoying a revival. Scrubbing the wood with a wire brush removes the soft material between whole cellulose tubes and a lime solution is then applied which deposits in these areas. The alkaline lime solution reacts with the tannic acid in oak turning it grey. Safer proprietary preparations now use lime in a medium of paraffin wax. All surface scratches have to be removed before this treatment with 220–240 grade sandpaper, as liming will show up every defect. Any preparation should be applied with a cloth in a circular motion and the surface can be buffed up after 40 minutes.

All finishes other than lining are also relevant to plywood. If plywood is used externally the correct grade should be used. It should be made with

weatherproof adhesive and of durable grades of timber veneer, e.g. gaboon or African mahogany (generally hardwoods, not softwoods).

A full list of natural finishes for wood is given in Table 3.7.

Figures 3.7, 3.8 and 3.9 show a variety of solutions to the problem of maintaining the surface of timber.

Metals

In order to coat metals and protect them successfully one needs an understanding of the mechanisms of corrosion. These mechanisms are outlined in MBS: *Materials* Chapter 9 and *Materials Technology* Chapter 6. On reflection, it is obvious that every metal in use in a building is a highly processed material which is in an artifically alloyed state. The natural state of all metals is their original form found in the Earth's crust. Metals may be oxides or sulphides but in their

Table 3.6 Natural finishes for internal and external wood

Application	Product type	System	Typical life to first maintenance*
External window joinery, doors and frames			
Softwood	Varnish Not recommended	–	–

	Exterior wood stain
	   Low build
	   Medium build
		2–3 coats	Variable according to product type but unlikely to exceed 3 years on full exposure

Hardwood	Varnish Exterior grade, full gloss	4 coats	Unlikely to exceed 3 years
	Exterior wood stain As for softwood	2–3 coats	As for softwood
Plywood, e.g. door panels	Varnish Not recommended	–	–
	Exterior wood stain Low build	2–3 coats	As for softwood. Salt-staining possible
External boarding, cladding, bargeboards, soffits, fascias			
Softwood	Varnish Not recommended	–	–

	Exterior wood stain
	   As for window joinery
	   Also zero build†
		2–3 coats	As for window joinery

Hardwood	Varnish Exterior grade, full gloss	4 coats	As for window joinery
	Exterior wood stain As for window joinery	2–3 coats	As for window joinery
Plywood	Varnish Not recommended	–	–
	Exterior wood stain As for window joinery	2–3 coats	As for window joinery. Salt-staining possible
Gates, fences, handrails			
Softwood	Varnish Not recommended	–	–

	Exterior wood stain
	   As for window joinery
	   Zero build†
		2–3 coats	As for window joinery

Hardwood	Varnish Not recommended	–	–

	Exterior wood stain
	   As for window joinery
	   Zero build†
		2–3 coats	According to required appearance

Internal general joinery, surfaces, linings and fitments			

Softwood, hardwood, plywood‡	Varnish (over decorative wood stain if required)
	   Full gloss or mid-sheen (semi-gloss)
	   Polyurethane, two-pack or moisture-curing for exceptional abrasion resistance
		2–3 coats	Variable according to type and service conditions but typically up to 5 years in 'average wear' environments

	Wood stain Some exterior wood stains may be suitable for interior use but refer to manufacturer's recommendations	1–2 coats	Variable according to type and service conditions. May give lifetime service in some situations

Note: Reference should also be made to BS 6952: Part 1 for guidance concerning the classification and selection of exterior wood coating systems.

* Life expectancies shown assume application to dry, sound timber which, if necessary, has received preservative treatment and are based on performance in 'moderate' environments.

† Zero build wood stains should only be employed on rough sawn timber.

‡ When a high standard of finish is required on internal hardwood surfaces, special-purpose wood finishes, e.g. French polish or lacquer, are generally used.

Source: Based on Table 12 BS 6150:1991

Table 3.7 Finishes for wood

Description	General composition	Characteristics and usage
Varnish, exterior grade, full gloss	Typically, drying-oil/phenolic or alkyd resin	This provides a tough, water-resistant coating, used principally as a clear protective finish for exterior hardwood. May need frequent maintenance. Tendency to lose flexibility with age
Varnish, interior grade, full gloss	Typically, drying-oil/alkyd, urethane or urethane/alkyd resin	Harder than exterior or grade type and is more suitable for use on interior hardwood and softwood joinery. Some types may be sufficiently abrasion-resistant to be suitable for use on hardwood and softwood floors, counter tops and similar 'hard wear' locations
Varnish, eggshell, satin or matt finish	Composition generally as for varnish above but adjusted to provide a lower level of gloss	Generally as for interior grade but is likely to be less suitable for use in 'hard wear' locations
Varnish, polyurethane, two-pack or moisture-curing one-pack	Two-pack types are supplied as separate base and 'activator' which are mixed before use to initiate chemical curing. With one-pack moisture-curing types, the reaction is initiated by absorption of moisture from the atmosphere or from the surface to which the material is applied	These coatings provide extremely hard, strong films with exceptional resistance to abrasion. The stresses set up within the film may lead to peeling and flaking especially on exterior woodwork whose surface, through long exposure without protection, has become degraded. In general, the use of this type of coating is best confined to interior woodwork where exceptional resistance to abrasion and possible chemical attack is required
Interior wood stain	Coloured pigments or dyes, with or without drying oil, emulsion or solution binders	Used essentially to modify or enhance the appearance of interior wood without obscuring its grain and is usually overcoated with clear finishes
Alkyd paint	See Table 5.1 of BS 6150	Traditional wood finish. Available in gloss, semi-gloss and matt finishes. Tendency to lose flexibility with age
Exterior wood stain, semi-transparent, zero build	Resin solution of low viscosity with fungicide and pigment	Water-repellent penetrating stain suitable for brush application. Imparts little or no sheen to surface of wood. Because of very low film thickness, offers little resistance to passage of water vapour and, in consequence, moisture content of wood may fluctuate considerably. Stains of this type can be used on interior woodwork but before doing so, it should be ascertained from the manufacturer that the fungicide contained does not constitute a health hazard
Exterior wood stain, semi-transparent, low build	Solvent-borne or water-borne resin solution with pigmentation and fungicide	Because of its higher resin content, this type of stain will normally impart a noticeable sheen to the surface. It is less penetrative than zero build, above, and offers greater resistance to water vapour movement, so fluctuations in the moisture content of the wood are less pronounced
Exterior wood stain, semi-transparent, medium build	Generally as for low build above but higher resin content	Because of its higher resin content, this type of stain will normally impart a noticeable sheen to the surface. It has higher build than low build and offers greater resistance to water vapour movement, so fluctuations in the moisture content of the wood are less pronounced
Exterior wood stain, opaque, medium build	As for low build above	This type may be regarded as intermediate between a stain and a paint. It has low gloss, but the texture of the wood remains evident because of the differences in penetration within the growth rings. Some opaque stains may be used over weathered but sound paintwork
Special wood finish systems	As for low build above	Multi-coat systems of improved performance on wood. For optimum service the full system should be applied to bare wood

Source: Based on Table 6 BS 6150:1991

Figure 3.7 Pope House now relocated at Woodlawn Plantation, Virginia, by Frank Lloyd Wright, 1940. One of the Usonian Houses, showing careful choice of materials so that applied finishes are minimized or avoided altogether. The timber is cypress wood from Florida.

Figure 3.8 Fishermen's drying sheds at Hastings, East Sussex. Maintenance of these structures was by a mixture of creosoting and straightforward patching with new timber when needed. Restoration between 1985 and 1991 changed the ethos.

natural state they are stable compounds. After refinement metals are unstable materials and actively seek to re-combine with oxygen or other elements. We call this process *corrosion*.

The most stable coatings are those that can form an oxidized layer on the surface. That layer should be thick and even to prevent further corrosion. Paint coatings only work because they have a thickness which prevents oxygen from reaching the metal layer

Figure 3.9 (right) Timber fence post detail. As timber ages softer material is eroded creating greater surface area for exposure to ultra-violet light and cavities for water penetration. Consequently deterioration becomes faster as crevices provide opportunities for infestation. Painting on this surface would be unthinkable. This degree of surface irregularity does exist on badly or unprepared timber. It is visible under a microscope and affects the adhesion of paint which is only microns thick (10^{-6} mm).

within an acceptable period or they combine with the metal in such a way as to form a new chemically inert coating. Whatever mechanism is used to protect metal from decay it must be applied quickly before extensive corrosion products form on the parent metal (otherwise coatings are adhering to these weaker deposits and not to the substrate). Consequently the cleaning of metallic surfaces is not aimed at removing foreign debris but primarily performed in order to remove initial corrosion and to inhibit temporarily the formation of any new deposit. Cleaning can be done mechanically (blast cleaning) or chemically (pickling and etching).

Inspection In the application of coatings to metals care should be taken that the work is checked by an inspector who is independent of the contractor carrying out the work. This is standard practice in America and is also recommended in BS 5493:1977 section 4. ASTM D3276:1986[4] gives a checklist of applications with a very detailed account of methods of cleaning metals. This is paralleled by CP 3012:1972 which gives very detailed methods for cleaning all types of metals. There are guidelines for making observations at particular stages of the coating process with reference to tests that can be undertaken on site to check quality. This is to ensure longevity of the film with particular regard to the repercussions of failure. Corrosion fatigue will often have its origins in the failure of coatings.

Testing For example in ASTM D3359:1993[5] failure of adhesion in coatings is measured in a tape test (method A): a cross cut is made (a figure X). The tape is placed carefully over the X and then pulled away. The degree of film material that comes with the tape and its position relative to the cut indicates the grade of adhesion. In standard D3363:1992[6] film hardness can be checked on site testing with pencils of known grades of lead (from 6H to 6B). The pencil is pressed into the surface until a grade is found that will not mark the paint. This determines the relative softness of the film and hence the degree of hardness. These standards must be read carefully as the instructions for making the tests are very precise. The principles outlined above are only valid if testing is carried out in the exact manner described in the standard. Otherwise comparative values are invalid.

Iron and steel (ferrous metals) BS 5493:1984 is a comprehensive code of practice for the *Protective coating of iron and steel structures against corrosion*. It is a complete guide for the design, specification,

Figure 3.10 Flaking. Les Halles, Paris, 1986. Inadequate preparation of the galvanized substrate has resulted in loss of adhesion and embrittlement of the coating.

inspection and maintenance of coatings systems and is a fundamental reference book.

The preparation of steel surfaces is critical (Fig. 3.10) and, however good the surface cleaning to remove impurities, initial corrosion, etc., initial protection must be carried out as soon as possible and certainly not later than four hours after cleaning. Otherwise surface coatings will be adhering to an oxide layer, and not to the substrate at all. All steel is best treated in controlled conditions before delivery to site. For the surface preparation of steel see section three of BS 5493:1977. See also Swedish Standard SFS 055900 *Rust grades for steel surfaces and preparation grades prior to protective coating*.

General preparation Solvents are used to dissolve oil and grease; surfaces should always be well rinsed after application. *Alkali cleaning* will take off old paint and if used, surfaces should also be washed with water after application. *Wire brushing* and scraping with metal tools is a fairly crude and unsatisfactory method for preparing metal surfaces. *Blast cleaning* is more effective in removing dirt and corrosion. Metal shot or varying sizes of sand or grit in a stream of air or water may be used. Blast cleaning will also give a surface topography that improves adhesion. (See BS 5493:1977 for further advice and specification on blast cleaning). *Pickling* and *acid etching* will remove scale. *Flame cleaning* is used to evaporate moisture from the face of metals and can help to detach light corrosive deposits but it should be used with caution. It could affect the strength of metals locally by raising metals to their T_m (melting point) and possibly initiating re-crystallization. Loss of strength is probable especially metal sections thinner than 5 mm. Priming can be carried out on metal that is still warm to the touch, but not hot.

Atmospheric parameters   All metal surfaces must be dry. Any moisture on the face of the metal, if trapped, can initiate corrosion below the paint film and cause intergranular corrosion. This can cause the paint to flake off and spalling of finishes generally, leading to pitting in the surface of the material. To avoid condensation on the face of the metal, the temperature of the metal must be above the air dewpoint. It is advisable only to carry out surface coating if the air temperature is above 5 °C and the relative humidity is below 80%.

Priming   Priming should follow quickly after surface preparation. There is general agreement in Britain and the USA that this should happen within four hours of preparation. It is recommended that priming should be carried out within one hour of surface preparation on metal that has been blast cleaned, because the surface has then a fine micro-topography with a greater surface area available for oxidation.

The priming coat must be good enough to resist normal weather conditions for some time. Steel may stay exposed in a primed condition because frame structures are often erected before finishing coats are applied.

Primers for metal are listed in Table 3.8. Paint systems for iron and steel are listed in Table 3.9.

Coatings on wet-trade finishes

There are some general points to be made about plaster, cements, concrete and brickwork mortar. All these materials set by the mechanism of hydration, i.e. the chemical reaction with water which generates new compounds. Rates of hydration are slow and cannot be hurried. There is no point applying heat to 'dry out' the fabric of a new building. This will take away moisture that is needed to assist the process of hydration and will leave the surfaces of materials more friable and

Table 3.8 Primers for metal

Description	General composition	Characteristics and usage
Pretreatment, wash or etching primer, two-pack	Typically, polyvinyl butyral resin solution with phosphoric acid (as separate component) and zinc tetroxychromate pigment	The main function of these primers is to improve adhesion of paint systems to non-ferrous metals. They may also be used to provide temporary protection to blast-cleaned steel and sprayed-metal coatings. The two-pack types generally give superior performance but may be less convenient to use. Application of primers of this type does not usually obviate the need for the application of a normal type of primer subsequently. Most pretreatment primers may be used in conjunction with conventional and specialist coating systems. *Colour*: Typically low-opacity yellow (two-pack types) or blue (one-pack types)
Pretreatment, wash or etching primer, one-pack	Typically, polyvinyl butyral/phenolic resin solution with tinting pigment	
Red lead primer to BS 2523: type A	Linseed drying-oil type binder with red lead (types A and B) as the sole pigments	These are primers of traditional type for iron and steel especially in new construction when there is likely to be a lengthy delay between erection of the steel and completion of painting. More tolerant of indifferent surface preparation than most other metal primers, they are slow in hardening which is disadvantageous when early handling or recoating is necessary. *Colour*: Orange red
Red lead primer to BS 2523: type B		
Red lead/red oxide primer	Typically, drying-oil/resin type binder with red lead/red oxide pigmentation	These primers for iron and steel are quicker in drying and hardening than linseed oil/red lead primers and are therefore more suitable for use in maintenance work or when early handling or recoating is necessary, although they are of lower durability on exposure without top coats. *Colour*: Red-brown

Source: Based on Table 2 BS 6150:1991

Table 3.9 Paint systems for iron and steel

Surface condition, primer	Finishing system	Total film thickness (μm)	Typical life to first maintenance in environments indicated
Blast-cleaned Two coats oil-primer	One coat oil-based undercoat and two coats alkyd gloss finish or two undercoats and one coat of gloss finish* Two coats micaceous iron oxide paint†	170–205	*Exterior 'moderate'* Up to 10 years *Interior 'moderate'* Over 10 years
	Two coats aluminium paint One coat oil-based undercoat and one coat alkyd gloss finish or two coats of gloss finish Two coats micaceous iron oxide paint†	115–45	*Exterior 'moderate'* Up to 5 years for film thicknesses exceeding 125 μm *Interior 'moderate'* Up to 10 years
Manually cleaned Two coats red-lead primer as to BS 2523, zinc phosphate‡	Two coats aluminium paint One coat oil-based undercoat and one coat alkyd gloss finish or two coats of gloss finish Two coats micaceous iron oxide paint†	125–50	*Exterior 'moderate'* Up to 5 years *Interior 'moderate'* Up to 10 years
One coat primer as above	Two coats aluminium paint One coat oil-based undercoat and one coat alkyd gloss finish or two coats of gloss finish Two coats micaceous iron oxide paint†	85–115	*Exterior* Not recommended *Interior 'moderate'* Up to 5 years

* Mid-sheen alkyd finish may be used in 'mild' interior environments.
† MIO paint may be covered with alkyd gloss finish (if manufacturer approves) to provide specific colour.
‡ Zinc rich primers are also used but advice should be sought regarding compatible top coats; problems can arise with many alkyd or oil-based systems.

Source: Based on Table 13 BS 6150:1991

more likely to perform badly.

An approximate time for proper hydration is about five weeks for 25 mm of wet construction. Full-set plasterwork should wait a further three weeks before decoration begins. Screeds of 50 mm in thickness will take ten weeks to set fully. See BRE Digest 163 for advice about drying out buildings. Moisture meters or hygrometers should be used to check the moisture content and humidity of the construction before finishing coats are contemplated.

The materials used in wet construction often contain salt impurities which re-crystallize and are seen as efflorescence at the surface. Depending on the salts present, either hard glassy skins will form (potassium sulphate), or the more recognizable white fluffy compounds (sodium sulphate, magnesium sulphate). Any efflorescence is likely to damage new coatings, so it should be removed and the surface brushed down every few days until the reaction ceases.

Any form of construction using Portland cement or lime will be alkaline in nature and care should be taken that finishing materials are compatible. A 'saponification' reaction can occur with oil compounds in coatings: paint films effectively turn into soap. The effect is not quite so literal, but oil films will soften as a result. Alkali-resistant primers must be specified if oil-based coatings are to be used. To avoid this problem entirely, use only finishing coatings that are designed for these kinds of backgrounds and are classed clearly as *masonry paints*. Tables 3.10 and 3.11 list the kinds of paint systems available for internal plaster, concrete, brick, block and stone; and external renderings, concrete, brick, block and stone.

Plaster   Ideally all plasterwork should be fully dry before applying any paint coatings. Most paint films are compatible with plaster and the specification of paint finishes (as against wallpaper) is increasing.

Table 3.10 Paint systems for internal plaster, concrete, brick, block and stone

Substrate condition	Finish type	Primer	Finish system	Typical life to first maintenance*
Dry (r.h. below 75%)†	Alkyd gloss, mid-sheen or matt	Alkali-resisting primer or, plaster only, water-thinned primer	*Gloss finish* One coat oil-based or emulsion undercoat; one coat alkyd gloss finish	5 years or more
			Mid-sheen finish Two coats alkyd mid-sheen finish	5 years or more
			Matt-finish Two coats alkyd matt finish	Up to 5 years
	Emulsion paint	Primer not usually required. A well-thinned first coat of emulsion paint may be required on surfaces of high or variable porosity	*Matt or mid-sheen finish* Two or three coats general purpose emulsion paint, matt or mid-sheen [1]	5 years or more
			Matt, high-opacity finish Two coats 'contract' emulsion paint. One coat, spray-applied, may suffice in some situations [2]	Up to 5 years
	Multi-colour	Primer or basecoat as recommended by manufacturer	*Multi-colour finish* Usually one coat spray-applied but refer to manufacturer's instructions [3]	10 years or more
	Textured	Primer not usually required but refer to manufacturer's instructions	*'Plastic' texture paint* Normally one coat but may require over-painting [4]	Indefinite in environments in which normally used but likely to require periodic over-painting to maintain appearance
			Emulsion-based masonry paint, heavy-texture Normally one coat but refer to manufacturer's instructions [5]	
	Masonry paint, mineral type	Check with manufacturer	Check with manufacturer	Indefinite
	Cement paint (not on gypsum plaster)	Primer not required	*Cement paint* One or two coats [6]	5 years or more in situations for which cement paint is generally used

continued

There are significant differences between different types of plaster that may affect the performance of coatings. Using BS 1191:1973 as a classification for plasters, gypsum plasters are classified into four categories A, B, C and D. *Grade A* is pure plaster of Paris and can be treated as *Grade B*. Both are 'soft' plasters with good but variable absorption. Over-wetting is not recommended. *Grade C* plasters are

Table 3.10 continued

Substrate condition	Finish type	Primer	Finish system	Typical life to first maintenance*
Drying (some damp patches may be evident, r.h. 75–90%)	Emulsion paint	As for dry substrates	As [1] and [2] above	Generally as for similar systems on dry substrates but some risk of failure at higher moisture levels
	Multi-colour finishes possible but consult manufacturer	As for dry substrates	As [3] above	
	Textured paints possible but consult manufacturer	As for dry substrates	As [4] and [5] above. If over-coating is necessary, emulsion paint should be used	
	Masonry paint, mineral type	Check with manufacturer	Check with manufacturer	Indefinite
	Cement paint (not on gypsum plaster)	Primer not required	As [6] above	As for dry substrates
Damp (obvious damp patches, r.h. 90–100%)	Emulsion paint possible	Primer not recommended	As [1] and [2] above. 'Contract' types are usually more permeable than general purpose types and less prone to failure on damp substrates	High risk of early failure
	Masonry paint, mineral type	Check with manufacturer	Check with manufacturer	Indefinite
	Cement paint (not on gypsum plaster)	Primer not required	As [6] above	As for dry substrates
Wet (moisture visible on surface, r.h. 100%)	Cement paint (not on gypsum plaster)	Primer not required	As [6] above	Generally as for dry substrates but some risk of failure

* Life expectancies shown assume application to dry, sound substrates, qualified as indicated for other substrate conditions, and are based on performance in 'moderate' internal environments.
† 'r.h.' refers to the relative humidity in equilibrium with the surface.

Source: Based on Table 15 BS 6150:1991

harder, but if they have been artifically dried before hardening there will be particles of plaster available for chemical reaction once water is applied, which may cause expansion and blistering. They are also very slightly alkaline which should not prove a problem unless lime has also been added to increase workability. *Grade D* (Keene's Cement) is the hardest plaster available and the surface can be so dense that adhesion of paints can prove difficult. Keene's cement is used on squash courts, for example. It has been the practice to prime this particular plaster 'following the trowel', i.e. immediately after the initial set, usually within three hours. The primer used is well thinned with only a small amount of oil and is a traditional and not a modern type. Unfortunately not all classes of plaster specified in the British Standards are now readily available.

External renders (Portland cement)   Repairs to old renders should be carried out in accordance with BS 5262:1991. Fig. 3.11 shows typical cracking of paint on external render.

Table 3.11 Paint systems for external plaster, concrete, brick, block and stone

Substrate condition	Finish type	Primer	Finish system	Typical life to first maintenance*
Dry (r.h. below 75%)†	Alkyd gloss and modified alkyd gloss	Alkali-resisting primer	One coat oil-based undercoat; one or two coats alkyd gloss finish	3–5 years or more
	Emulsion paint, general purpose if suitable for external use	Primer not usually required	Two coats general purpose emulsion paint [1]	Up to 5 years
	Masonry paints, solvent-thinned	Alkali-resisting primer or as recommended by manufacturer	*Smooth or fine-textured types, solvent-thinned* two coats [2]	5 years or more
			Thick, textured types, solvent-thinned Usually one or two coats applied by spray, often by specialist applicators [3]	10 years or more
	Masonry paints, emulsion-based	Primer not usually required	*Smooth or fine-textured types, emulsion-based* Two coats [4]	5 years or more
		Primer not usually required but refer to manufacturer's recommendations	*Heavy-textured types, emulsion-based* Usually one coat applied by roller [5]	10 years or more
	Masonry paint, mineral type	Check with manufacturer	Check with manufacturer	Indefinite
	Cement paint	Primer not required	Two coats cement paint	Up to 5 years
Drying (some damp patches may be visible, r.h. 75–90%)	Emulsion paint, general purpose	As for dry substrates	As [2] above	Potentially as for dry substrates but some risk of earlier failure at higher moisture levels
	Masonry paints, emulsion-based	As for dry substrates	As [4] or [5] above	
	Masonry paint, mineral type	Check with manufacturer	Check with manufacturer	Indefinite
	Possibly solvent-thinned masonry paints but refer to manufacturer's recommendations	As for dry substrates	As [2] or [3] above	Potentially as for dry substrates but some risk of earlier failure at higher moisture levels

continued

Most external renders are alkaline due to their content of Portland cement and sometimes include integral lime constituents. If these renders are new, and not fully hydrated a saponification reaction can occur. Prime with an alkali-resisting primer or use a base coat that has been recommended by the paint manufacturer. Finish with alkyds, emulsions, masonry paints or cement paints.

Table 3.11 continued

Substrate condition	Finish type	Primer	Finish system	Typical life to first maintenance*
Damp (obvious damp patches, r.g. 90–100%)	Masonry paint, mineral type	Check with manufacturer	Check with manufacturer	Indefinite
	Cement paint	Primer not required	two coats cement paint	As for dry substrates
	Possibly emulsion-based masonry paints but refer to manufacturer's recommendations	Primer not recommended	As [4] or [5] above	Potentially as for dry substrates but high risk of earlier failure
Wet (moisture visible on surface, r.h. 100%)	Cement paint	Primer not required	two coats cement paint	As for dry substrates but some risk of earlier failure

* Life expectancies shown assume application to dry, sound substrates, qualified as indicated for other substrate conditions, and are based on performance in 'moderate' external conditions.
† 'r.h.' refers to the relative humidity in equilibrium with the surface.

Source: Based on Table 16 BS 6150:1991

Alternative coatings   There is a revival in the use of some old finishes. These are labour-intensive to mix up and apply, but very cheap and suitable for all ages of building and particularly for older properties.

Limewashes   Care must be taken in the use of lime; skin should be protected from contact and masks worn to prevent inhalation. Quicklime and tallow, putty lime and linseed oil or baglime can all be used. Putty lime is thought to be the longest lasting. After mixing with water, pigments are added. In addition to powdered pigments, acrylic colours can also be used. For more information see SPAB Information Sheet No. 1 (Jane Schofield, 1991).

   For conservation use English Heritage and the National Trust recommend a proprietary polymeric limewash, which is a hydrate of lime with microscopic glassy spheres, and a polymer binder.

Concrete   The painting of concrete should be avoided. Concrete has a durable exterior if properly specified and painting can sometimes lead to the

Figure 3.11 (left)   Flaking. Finish on external render. There are a number of serparate checking patterns that mirror the direction of induced stress (top left). As the paint film ages and hardens adhesion is lost and the film cracks in a characteristic hexagonal pattern typical of fracture in sheets. Note how the generation of the large fractures in the surface coating follows cracking in the render below.

harmful entrapment of moisture. There is some movement of free water in and out of the gel pores, and if this is inhibited using paint films that are permeable, blisters can occur. However, in order to give protection to re-inforcement (steel rods) in the body of concrete that may be only lightly embedded (especially on existing old concrete) films have to be impermeable and it is recommended to use chlorinated rubber and bituminous coatings. One needs to protect the re-inforcement, but this may harm the concrete. This is a dilemma, and one that highlights the problems in coatings generally.

The surface of concrete first has to be properly cleaned before coating. Brush cleaning will remove dust and debris initially, but air-blast cleaning at between 80 and 100 pounds per square inch pressure should be used 600 mm from the surface to remove all surface dusting. It should then be cleaned with water, and then, if necessary, with detergents or steam cleaning (which is often thought more effective) to remove any oil or grease.

If the purpose of painting is to improve an existing concrete surface, it is worth investigating two other methods first. The first one is to abrade the surface mechanically or to etch with acid the material to gain a different surface roughness or texture. Acid should only be used after the surface has been cleaned to a certain standard (ASTM D4258:1992).[7] Acid-etching uses different concentrations of hydrochloric, sulphuric or phosphoric acids. The concentration of acid varied according to the depth of etching required. After application the surface can be brushed down to give textures that vary from fine to coarse grades of sandpaper. It is worth mentioning that the less smooth the surface topography the greater the likelihood of entrapment of dust and grime, which can harm the surface further in the future.

Brickwork and stonework   It is not advisable to paint either brickwork or stonework. They are usually durable-enough materials and any coatings will often create new problems. There should be a free movement of moisture through a masonry structure and paint can do a great deal to inhibit the passage of water vapour which can become entrapped. Because of the degree of porosity, especially in brickwork, frost action can be more noticeable on externally painted walls, and free moisture can also encourage latent efflorescence. In the case of sedimentary stone this can be evident as 'subflorescence'.

Before painting concrete, cements, brickwork or stonework, a full investigation should be made as to whether the surface can be cleaned, mechanically or

chemically. Often paint appears to be the first 'cost-effective' solution to improving the look of an external surface. But the problems caused by applying paints indiscriminately can cost far more in terms of new defects and repeated maintenance in the long term than careful and effective cleaning. For specialist advice in restoration contact the Society for the Protection of Ancient Buildings. SPAB is at 37 Spital Square, London EC1 and publish a range of material helpful in understanding the detailing and restoration of ancient buildings.

Architectural wall coatings   These coatings are similar to industrial floor finishes. They are hard finishes and use two-pack resin systems such as epoxy-polyamides, or polyester-epoxy coatings. These are now widely used. These particular finishes are a viable alternative to tiling in some public areas, as they have a lower initial cost and require little maintenance. They show good adhesion, are chemically resistant and can be scrubbed clean. Fire resistance may be a problem and checks should be made on flame-spread.

Other substrates   The substrates dealt with so far cover the range of basic materials used in a building. There are a number of other materials, such as composite boards, hardboards, paper, fabrics and plastics which will need a particular paint specification. It is best to take the manufacturers' advice when choosing a paint system. The problems in applying new paint systems to backgrounds which have previously been coated with bituminous coatings (especially old timbers) are well-known. Aluminium primers should be used in this instance. For other general advice BS 6150:1991 should be the first source of reference. With any unusual material, particularly composite materials, the most important decision is in the priming system. This should take into account the degree of absorption of the material as well as chemical compatibility. Advice is given more specifically on priming these materials, in Table 3.12.

3.6 Adhesives

Introduction

Adhesive bonding can be divided into two main categories: 'structural bonding' exceeds $10 \, \text{N/mm}^2$ in tensile strength and weaker bonding is 'non-structural'. Adhesives are also categorized according to durability, denoted by the following letters:

  MR   Moderately weather-resistant (to several years exposure)

Table 3.12 Site priming of fibre building-board, wood chipboard and plasterboard

Type of board or sheet	Primers	
	With alkyd and oil-based finishes	With water-thinned finishes*
Hardboard, mediumboard, medium density fibreboard (MDF) and softboard	Primer-sealer Water-thinned primer Aluminium wood primer	Not usually required, but, for absorbent board, first coats may need additional thinning
As above, flame retardant treated	Alkali-resisting primer	Alkali-resisting primer
Softboard, bitumen-impregnated	Aluminium wood primer	Not usually required
Wood chipboard	Oil-based wood primer Primer-sealer Water-thinned primer Aluminium wood primer	Not usually required except possibly that oil-based primer is recommended for single layer boards to prevent swelling of chips
Plasterboard	Primer-sealer Water-thinned primer	Not usually required

* Water-thinned multi-coilour finishes are usually applied over a special base coat.

Source: Based on Table 17 BS 6150:1991

BR   Water- and weather-resistant, withstanding temperatures up to 100 °C and attacks by organisms

BWP   Boil-proof and weather-proof. Long-term durability.

Adhesives are either natural or synthetic. It may be important to establish this at an early stage if specifiers are keen to comply with the EC directives on VOC's (volatile organic compounds). The hazard must now be identified not just for operatives in the workplace, but for the population in general over an extremely long timescale.

The most common synthetic resins are:

Acrylic resins, which include cyanocrylates and two-part adhesives which use a resin with a catalyst

Modified phenolic resins which need heat and pressure for curing

Epoxy resins usually supplied with a hardening compound

Polyurethanes

Plastisols which are modified polyvinyl/pvc resins hardened by heat

PVA (polyvinyl acetate) types for wood

They all have a variable strength, although a maximum of about 20 N/mm^2 can be achieved in most cases.

In surface preparation the thickness of the bond should be minimized, to ideally between 0.05 and 0.15 mm. Adhesive joints which are continuous are much more efficient as the stress is spread. Rivets and other isolated fixings may seem efficient, but they actually localize stress at the point where material failure becomes critical.

Substitutes for organically based adhesives

As part of the EC's attempt to limit VOC's, research in Germany is currently exploring using natural polymers as adhesives for particleboards and plywoods. Formaldehyde emissions would be reduced, and these materials will come from renewable resources. So far, it appears that natural polymers must still be used in conjunction with di-isocyanates, as natural polymers do not develop sufficient bond strength on their own. One solution may be the development of different production techniques.[8] Wood-based panel products use synthetics: condensates of phenol, resorcinol, urea and melamine with formaldehyde. Adhesives can be made from tanninsulfonates (from the waste products of conifer bark) as a substitute for half of the resorcinol normally used. An important point was made by Kreibich and Hemingway:[9] 'Declining tree quality has increased the use of adhesives in the manufacture of large, strong, structural materials from wood.' It is almost certain that more laminated wood products will use more adhesives. Perhaps effort should be directed to finding the causes for poor tree quality and improving it, rather than relying on expensive and more damaging chemical engineering.

Table 3.13 Polymer flooring applications

Type of polymer	Method of laying	Size and indicative price per square metre
Mastic asphalt Natural or manufactured bitumen with aggregates, fillers and sometimes pigments Black, red or green BS 6577:1985 *Specification for mastic asphalt building* BS 6925:1988 *Aggregates* BS 8204:1988 Sections 4–6 Flooring Handbook, Mastic Asphalt Council	Laid in hot liquid state, and floated to level finish. If in two layers pour over separating medium of sheathing felt	Continuous material, but thicknesses are 15–50 mm depending on grade required for application Price: £15–18 for 20 mm thick
Pitch Mastic Differs from asphalt by having coal tar pitch as the binder for aggregates BS 5902 BS 8204:1988 Sections 5 and 6	As mastic asphalt	Increase on mastic asphalt thicknesses for the same application
Rubber Can be smooth or textured with ribs, or raised studs, supplied as a sheet material or as tiles Great range of colours BS 1711:1991 *Solid rubber flooring* BS 3187:1991 *Electrically conducting rubber flooring* BS 8204:1988	Bond with recommended adhesive, usually epoxy or neoprene and apply pressure	Sheets are 910, 1370 or 1830 mm widths with thicknesses 3.8–6.4 mm for domestic use and up to 12.7 mm in thickness for heavy industrial use or mats Tiles can be 229–914 mm square and from 6–13 mm in thickness Price: £26–30 for 5 mm thick
Vinyl (flexible) PVC binder with fillers, pigments and plasticizers to control flexibility, has thin or thick foam backing. Variety of colours and textures BS 3261:1991 Part 1 BS 8203:1987	Bond to base which can be self-levelling compound on screed, or overlay of hardboard or plywood over timber boards or direct on to chipboard and plywood flooring Material should be stored at 18 °C for 24 hours before laying	Sheet widths vary 1200–2100 mm and thicknesses 1.5–4.5 mm. Joints can be site welded for hygiene and continuity Tiles can be 225, 250 or 300 mm square, thicknesses 1.5–3 mm Price: £8.5–11 tiles; £12–15 sheet
Resin emulsions as binders with Portland cement Additional pigments and aggregates and fillers giving good range of colours, aggregates can be exposed BS 8204:1988 Section 6		Thickness 4–12 mm Price: £4–7 for 3 mm coat; £7–15 for 5 mm coat

Polymers and recycling

There is sufficient volume of waste in the plastics industry to generate recycling programmes that can supply significant quantities of materials for new products. To assist recycling there are now labelling schemes to assist re-processing. BS 7746:1994 *Generic identification and marking of plastics products* 'specifies a system for the uniform marking of products that have been fabricated from plastics to help identify plastic products for subsequent decisions relating to handling waste recovery or disposal.'

3.7 Floorings

Polymers have wide uses in the flooring industry as the composition of tiles and sheets can be engineered to suit almost any situation. Resins can be used to give seamless tough finishes, or used as sealants on other floor finishes. Because of the chemical composition of polymer-based flooring systems, surface preparation and the correct use of adhesives is critical and manufacturers' advice and specifications should be followed. All polymer-based flooring materials are best washed with warm water and neutral detergents.

Table 3.14 Resin flooring applications

Type of Polymer	Method of Laying	Size and indicative price per square metre
Polyester	On screed or power floated and well-finished slab. Thin layers need a good level base One-component pack sets by exposure to atmosphere	2–3 mm thick Price: £41–8 for 5–9 mm thick
Epoxy	On screed or power floated and well-finished slab. Thin layers need a good level base Liquid resin sets by mixing with a hardener	2 mm thick if sprayed or self-levelling, 3–6 mm thick if trowelled. Price: £18–24 if thin Price: £36–42 if thick
Polyurethane	On screed or slab. Can take up greater unevenness due to potential thickness	2–4.5 mm if self-levelling; up to 10 mm thick if trowelled Price: £??

Only the harder resins can withstand harder cleaning long-term. Most polymer-based floorings, especially rubbers and vinyls, can take a limited polish using liquid polishes dispersed as emulsions in water-based solutions.

Polymer floorings   Table 3.13 lists different types of polymer flooring.

Mastic asphalt is an elastic gel, in contrast to cement which is a rigid gel. (A gel is a mixture of solids in liquid.) Although the grading of aggregates may be similar in both cases, the matrix of the mastic has the ability to move under heat or pressure and so exhibits polymeric thermoplastic tendencies. Mastic asphalt is a heavy duty finish, resistant to water and alkalis but it can be broken down by oils and grease. It is affected by heat and will show marks from point loads but aggregates can be added to improve hardness and durability. Pitch mastic can be specified if greater resistance to oils and alkalis is needed. It will take a polish but like any flooring can be slippery if wet.

Vinyls are generally more resistant to oils and grease than rubbers.

Resin floorings   Resin-based flooring systems satisfy more stringent demands for hygiene in industry by providing seamless and continuous flooring that withstands most chemicals and can be thoroughly cleaned. Site work has to comply with latest EC and COSHH regulations to ensure that operatives are not exposed to high levels of fumes and that they wear protective clothing and in some cases breathing apparatus.

There are a great range of products which use resins in various combinations with pigments, fillers and aggregates (Table 3.14). Different resins will give slightly different performances. Epoxy resins have the highest resistance to water, acids, oils, alkalis and some solvents, but are more expensive, whereas polyester and polyurethane resins have more moderate resistance to alkalis and acids. For particular applications it is best to discuss specifications with the manufacturer.

Notes

1   Original directive 88/379 EEC, currently upgraded by Council Directive 93/18th April 1993. See also directive 67/548 EEC, upgraded by 92/30th April 1992 which relates to substances.
2   Test method for hiding power of paints by reflectometry.
3   Method for measuring moisture in concrete by the plastic sheet method.
4   Guide for paint inspectors (metal substrate).
5   Test methods for measuring adhesion by tape test.
6   Test for film hardness by pencil test.
7   Practice for surface cleaning concrete for coating.
8   Dix, Brigette and Marutzky, Rainer: *Adhesives from Renewable Sources* Chapter 17, Am Chem Soc 1989.
9   Kreicbich, Roland E. and Hemingway, Richard: *Adhesives from Renewable Resources*, Chapter 17: *Tannin-based adhesives for finger-jointing wood*, Am Chem Soc 1989.

4 Ceramic materials

4.1 Introduction

The term 'ceramics' covers a range of materials that all have a similar range of properties and behaviour, but may vary in their origin and appearance. Ceramics can cover natural as well as artificial materials, e.g. minerals, rocks, cement gels and glasses, as well as sintered clays which give us our traditional ceramics. Traditional ceramics are made from clays which are fired to give pottery, bricks and tiles. They are a combination of flint, feldspar and clay. Modern ceramics have the following characteristics:

They are *inorganic*
They have *non-metallic properties* (but metallic elements may be present)
They are *ionically* and *covalently bonded*
They often contain *a combination of metal and oxygen atoms* (or more generally gaseous and metallic elements, e.g. nitrogen can be substituted for oxygen in many alumino-silicates; they are then called 'sialons')

They are also often:

Hard
Good thermal insulators (good fire resistance)
Good electrical insulators
Chemically stable
Brittle
Of high compressive strength
Multi-phase materials

The ceramic materials used in building exhibit all these properties. The advantages of using many ceramic materials is that are already environmentally stable and will not oxidise further in the atmosphere. Unlike metals, which generally exist in an unstable state and must be protected, ceramics are unlikely to react with elements in the atmosphere and are normally used in their natural state, i.e. their body composition can be their finished skin. This makes them economic in terms of maintenance. Problems are more likely to occur when they are combined with other materials to give different properties. For example when concrete is used in combination with steel to improve its tensile strength by making a composite, carbonation and the subsequent corrosion of the re-inforcement will make the whole material unstable. Mechanical fixing details generally are also a source of failure. The basic ceramic material usually degrades slowly, and is a long-lived component, whereas the fixing mechanisms are often highly stressed and, if corrosion occurs, failure is dramatic.

Traditional ceramics

There are fired-clay products. Depending on the clay bodies used, the admixtures and the degree of firing, the phase microstructure of the ceramic can be manipulated.

Earthenware clays produce a soft porous material after firing at low temperatures. On fracture, the material shows cleavage around individual particles of clay. Glazed material shows a noticeable boundary between glazed and unglazed sections. It will have a Mohs' hardness of about 4 or 5 (see glossary section).

Stoneware clays are fired at higher temperatures and have a hardness of 7 or 8. If fractured, the material will break as a glassy body with a continuous fracture plane. A stoneware glaze appears to fuse with the clay body, which is extremely dense and non-porous. The silica in the stoneware forms a glassy matrix in which particles are suspended.

Porcelain is the most glass-like body, with the glaze appearing completely integral with the body. Although a fine-grained material, it is extremely tough and durable, as the microstructure of porcelain reveals needle-like crystals of *mullite* in a glassy phase matrix which prevents crack propagation.

Increases in the strength of ceramic materials can be achieved by minimizing glassy regions, through which crack propagation is dramatic, and also by manipulating the size and structure of crystals. Fractures require more energy to propagate through crystalline rather than glassy regions and if the crystals are small the work of fracture is higher. Even if there is only a partly crystalline region, as in porcelain, there is a toughening capability. The heat resistant ceramic hobs appear to be glassy but are a carefully controlled crystalline material. The material is seeded with nuclei for crystal growth and exact temperature control produces a completely crystalline material with strengths far exceeding that of porcelain. These materials are known as *glass ceramics* and their main features are high-temperature strength and resistance to thermal shock, due chiefly to the inclusion of high quartz phases. Even stronger glass ceramics can be made from the introduction of keatite crystals and other ranges are even machinable, their fracture toughness is so high.

New ceramics

The new ceramics, also sometimes referred to as technical ceramics, or engineering ceramics, have polycrystalline micro-structures and so have several components. Their purity is far higher than traditional ceramics, and instead of using raw clay mined directly from the ground, the components used are pure compounds with no unwanted elements or impurities. Powders are formed from these ingredients and are then cast, pressed, extruded or moulded into shape. These powders may also be set with organic binders. At this stage they may undergo machining before final sintering, sometimes in combination with additional applied pressure. It is the combination of pure materials and exacting production techniques that ensures the very high strength of these materials. ('Ceramics in a Metals World', James A. Spirakis in *Advanced Materials and Processes* 3/87.)

Because of their very high strength chemical bonding at an atomic level, pure ceramics can be used as fibres, whiskers, or as toughening particles in composite materials. To optimize these materials very specific elements are used which show this type of strong convalent bonding and also have very low atomic weights. Elements that fit this criterion are beryllium, boron, carbon and silicon which when

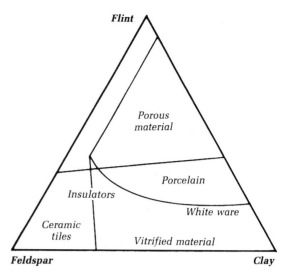

Figure 4.1 Traditional ceramics. Relation of materials to products obtained. (*After*: I.J. Mcholm 1983 *Ceramic Science for Materials Technologists* Leonard Hill, Blackie and Son.)

combined, as in boron carbide or silicon carbide, give an extremely strong group of materials.

A variety of traditional ceramics and their starting materials are shown in Fig. 4.1.

4.2 Glossary

Bonding   See section 4.3.

Calcination   Materials that have been calcined have been heated to a sufficiently high temperature to give off water and in some cases, especially in metallic ore extraction, carbon dioxide.

Cermets   This is a general term that relates to ceramic/metallic particle composites, but initially and more specifically to tungsten (W-Co) 'cemented' materials. In this instance the metallic phase forms the matrix, and is sintered from the powdered metal.

Clastic material   This refers to the construction of new stones from older material.

Faience   These are glazed *terra cotta* (earthenware) tiles used on buildings. The word, which is French, originates from 'Faenze' in Italy, which was historically a centre for the production of fired ceramics which were chiefly tin-glazed.

The composition from Mitchells Building Construction (1947) is given in Table 4.1.

Table 4.1 Composition of faience tiles

Compound	Percentage composition
Silica	75.2
Alumina	10.0
Ferric oxide	3.4
Calcium oxide	1.2
Magnesium oxide	(traces)
Alkalies and alkaline chlorides	0.5
Water	5.9
Organic matter	7.7

Glass A glass is a non-crystalline solid, or *amorphous solid*, as it has no regular structure. Glasses have the random order of molecules normally associated with liquids, which is why they are sometimes referred to as *super-cooled liquids*.

Mohs' scale of hardness This scale of hardness is based on the ability of a material to resist a scratch defect (Table 4.2). The tests should be made by hand and the examination for scratch deformation is made by eye. This method of testing is described in BS 6431 Part 13:1986 relating to ceramic floor and wall tiles.

Multiphase materials In solid materials there may be regions of different molecular structure. In many rocks these would be seen clearly as different mineral forms with individual, identifiable characteristics. Most ceramic materials could be called 'multiphase solids', as they are usually composed of particles with their own individual structures, set within a matrix which may be glassy but may also have its own identifiable structure.

Salts These are compounds resulting from the replacement of one or more hydrogen atoms in an acid by metal atoms, or electropositive radicals. They are usually crystalline at ordinary temperatures and include carbonates, chlorides, nitrates, phosphates, silicates and sulphates.

Table 4.2 Moh's scale of hardness

Scale	Material	Test
1	Talc	Can be scratched by fingernail
2	Gypsum	Can be scratched by fingernail
3	Calcite	Can be scratched by a copper coin
4	Fluorspar	Can be scratched by steel
5	Apatite	Can be scratched by steel
6	Felspar	Scratches steel with difficulty
7	Quartz	Scratches steel with ease
8	Topaz	Scratches steel with ease
9	Corundum	Scratches steel with ease
10	Diamond	Scratches steel with ease

Sintering Sintering is the fusing of particles together. The particles may be metallic, ceramic or a mixture of both. The powdered mixtures are heated until they fuse and coalesce with adjacent particles.

Materials in a powdered form have a very great surface area. This is important in fusion processes: the greater the surface area the more efficient the transfer of heat, and the faster and more controllable the process of fusion. Sintered bodies have characteristic gaps giving a degree of porosity. If these are filled, then a material must have been present in the manufacturing process that became a liquid phase.

Suction 'Suction' is a loose term. It is the 'property' of cementitious materials to take up water and other materials. This ability to absorb liquids is due to capillary action.

Terra cotta Terra cotta comes from the Latin for 'baked earth'. It is now used more loosely on a large scale for hollow ceramic fittings (such as chimney pots) which imitate stone. Terra cotta was used from the end of the nineteenth century. These fittings are made from graded clays often incorporating some re-ground material from earlier fired but substandard products.

Vitrification Materials that are fired at high temperatures and fuse to give a glassy state are said to be *vitrified*. They are supercooled liquids. Ceramic glazes vitrify on firing.

4.3 Bonding

Physical

Bonding is a term that has several different definitions according to which discipline is using the term. In building it is 'the securing of a bond between plaster and backing by physical means as opposed to mechanical keys' (Chambers Dictionary of Science and Technology). Mechanical keys or mechanical bondings refers to a physical locking of parts by their profiling. The physical means refer to a good take up by the backing material of the substance and good adhesion. Adhesion varies in its definition according to the scientific discipline concerned, but in physics is generally defined as:

'By intermolecular forces which hold matter together, particularly closely contiguous surfaces of adjacent media, e.g. a liquid in contact with a solid.' (Chambers Dictionary)

Adhesion is often achieved by a physical attraction through van der Waals forces between pore walls (usually displaying electronegativity from compounds containing oxygen) and the initial attraction with water molecules in solution (which are dipoles with positively charged hydrogen ions).

Chemical

In ceramics there are two systems of bonding, covalent and ionic. Covalent bonding is by far the strongest of all chemical bonds, involving the pairing of electrons shared between two atoms. The pairing structures try and attain the most stable of all configurations, i.e. that of the noble gases, which are the most stable elements. Consequently it requires great energy to split these bonds, which in normal materials processing is usually carried out by the application of heat.

Ionic bonding is the second strongest configuration and describes the transfer of electrons from a metal to a non-metal. The loss of electrons (negative particles) will make an element strongly positive, whereas the gain of electrons will make an element strongly electronegative. Consequently the ionic bond relies on strong electrostatic attraction between oppositely charged bodies. The final distribution of electrons also seeks to attain that of a noble gas.

A great deal of the bonding occurring in compounds is intermediate in nature between the two pure bonding types. The only other type of bonding, not found in ceramics, is metallic bonding where, although metallic atoms occupy regular positions in the formal geometry of the atomic lattice, the electrons act like a cloud and move freely through the lattice. This cloud interferes with wavelengths of light absorbing energy from light photons, but reflecting some, hence their opaque quality with some degree of lustre.

4.4 Fracture mechanics

There is a commonality about the methods of fracture in Ceramic materials. Most traditional building materials are brittle. This tendency towards a brittle state determines the method of fracture. Most traditional ceramics can display particular flaws which act as crack initiators. Even if these flaws are extremely small they can be effective as crack initiators because they concentrate stress in a small area. In particular ceramics which have small pores in the body of the material have built in stress concentrators, which will make the material predictably brittle. A.A. Griffith developed his famous theorem of crack propagation in the 1920s and his

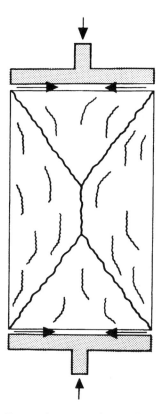

Figure 4.2   Fracture in compression specimens.

formulae (modified) are still used as a powerful model for the prediction of crack behaviour.

Fracture due to compression is shown in a variety of specimens in Fig. 4.2.

4.5 Frost and freezing mechanisms

Frost can have a devastating effect on buildings (Figs 4.3 and 4.4). Water expands on freezing by approximately 9%, and this is the major cause of the mechanical spalling of ceramic bodies. Ceramics absorb water and become saturated at the surface. If this saturated water freezes, it increases in volume, and this is sufficient to force materials apart. This simple explanation for 'frost damage' is now being further refined. Attention has now focused on the pore system of a material and the pressures that can be exerted within those pores, not just by the growth of ice crystals but by the pressurization of trapped chilled water which can be hydraulically rammed by adjacent crystal growth. This may be the real cause of mechanically applied pressure. Also, as ice crystals form, they continue to grow and are fed by adjacent

Figure 4.3 Brick arch, St Pancras Station, London. Extensive erosion of soft red bricks and stone is accelerated by frost action causing spalling of both materials, particularly of the brickwork. Note the dark encrustations of sooty material, forming a hard skin in sheltered regions. It is formed from a combination of salts, pollutants and water vapour held in this more sheltered area.

chilled water. So the mechanics of fracture may be due to prolonged crystal growth and not the volumetric expansion of water.

The size of pressure that can be exerted is inversely proportional to pore size and is related to pore size distribution. In turn, these also determine other properties which include strength, water absorption, saturation coefficients, rate of absorption and capillary size.

Tests to predict frost failure according to pore size and distribution have not been over successful, as they do not seem to be able to imitate the failure of the material in practice. The only way of establishing quality is still by use of freeze–thaw test cycles. The British Standard is BS 6431 Part 22:1968 *Method for determination of frost resistance*. The European standard EN 202:1991 describes how after water saturation, tiles are taken through a temperature cycle from $+15\,°C$ to $-15\,°C$ and back again. All sides of the tile are exposed to over 50 freeze–thaw cycles. Frost resistance is 'satisfactory' if tiles survive 15 freeze–thaw cycles. Destruction of ceramic materials usually starts from the glazed surface with fine crazing.

Figure 4.4 Brickwork, St Pancras Station, London. Frost action has damaged the surface of this brick; spalling patterns echo the irregularity of the inner pore structure of the material.

Rounded cracks then lead to a tear out of material, the rounded micro cracks transform in to shallow tips and the degradation is progressive. There is an increase in the surface porosity, a greater ability to take up water and the damage continues.[1] The early development of these cracks cannot be seen with the naked eye and the number of cycles of freezing is minimal compared with the cycles of testing (in the order of 10^2). The authors conclude that the integrity of the glaze coating is all important and the most important aspect to consider with regard to frost protection. In a paper published in 1986 Nakamura and Okuda[2] experimented with boiling clay roofing tiles in n-butanol for at least 24 hours to form a butoxy group within the outermost skin of the tiles, giving less water absorption and enhanced durability. A later paper by Romanaova[3] confirms the value of a strong intermediate layer between tile and glaze to improve frost resistance. In another paper, Nakamura[4] confirms several indirect evaluation factors which can influence frost susceptibility in clay roofing tiles. These are: water absorption, bending strength, capillary water saturation and pore size distribution.

Predictions of possible behaviour are difficult, but there is growing proof that denser tiles, i.e. those which are fully vitrified with a minimum pore diameter, may be at greater risk from frost damage. A pore with a greater diameter has room for expansion when water freezes, and is also able to allow evaporation of moisture.

Another variation of the hydraulic pressure theory describes how, in a highly porous material, the growth of ice crystals puts pressure on adjacent unfrozen water which may be unable to flow.

Instead of understanding frost action as the expansion on freezing of ice, if ice crystals grow with a source of unfrozen water still available then new crystal growth will put additional pressure on capillary walls (similar to frost heave in soils). This requires regions of coarse pores adjacent to microporous regions for this particular phenomenon to happen. Proof that this combination is critical to initiate frost damage can be seen in stone with this kind of pore distribution. Portland stone is prone to frost damage whereas sandstone with an overall coarse size of pore diameter is more resistant to frost damage.

A Finnish paper on arctic concrete technology describes a related observation.[5] Tests were carried out in extreme conditions using freeze–thaw tests in a range of $+20\,°C$ to $-60\,°C$. Concrete types tested included mixtures with and without air entrainment and also mixtures with hollow plastic microspheres. The concrete mixes that had been air entrained showed a better frost resistance. Concrete without air entrainment and with the microspheres showed an appreciable decrease in strength.

Most work on frost resistance has concentrated on tiling because of the continuous exposure in saturated conditions, to freeze–thaw cycles.

The following is a checklist for specifying ceramic materials for external use:

- The exact nature of the material, method of fixing and type of finish
- Degree of frost resistance
- Degree of tolerance to component dimensions, joint width and subsequent setting-out
- Maintenance

Notes

1 *Effect of the quality of glaze coating on the frost resistance of facade tiles.* Egerev, V.M. (Scientific Research Institute of Building Ceramics, USSR), Zotov, S.N., Romanova, G.P., Lykhina, N.S., *Glass Ceramics* V42, N. 7–8 Jul–Aug 1985, p. 351–372.

2 *Enhancement of frost durability by modifications of internal surface of clay roofing tiles.* Nakamura, M. and Okuda, S. in *Yogyo Kyokai Shi* V 94, N. 12, 1986, p. 1239–1242.

3 *Increasing frost resistance of facade glazed tiles.* Eger, V.M., Zotov, S.N., Romanaova, G.P. *Glass Ceram* V43, N. 1–2, Jan–Feb 1986, p. 66–68.

4 *Indirect evaluation of frost susceptibility of clay roofing tiles.* Nakamura, M., Mama, A., Matsumoto, S., Okuda, S., in *Yogyo Kyokai Shi* V94, N. 12, 1986, p. 1239–1242.

5 *Arctic Concrete Technology.* Kivekas, L., Huovinen, S., Hakkarainen, T. and Leivo, M. *Valt Tek Tutkimuskesku Tutkimuksia* N. 305, 1984. p. 149.

5 Applications of ceramic materials

5.1 Internal and external tiling

Historical uses of tiling are worth considering since ceramic technology has been available for thousands of years. The development of glazes on a solid skin finish to a building allowed for richness of decoration and for an applied finish that had a developed craftsmanship, whether by painting, a choice of inlaid marbles or selected mosaics. Even today, public arts projects often use ceramic techniques to change the scale of human interest in buildings. This finish is long lasting and can represent not just investment but a very particular attitude to design or colour prevalent at the time.

Although roof tiling has been used for centuries without much change, there have been considerable changes in the way people have used tiles as walling and flooring materials.

Early wall tiling (including mathematical tiles) is related to a localized vernacular and possible use of timber frame construction. The tiles were fastened with pegs onto battens fixed to the main frame. The joints of mathematical tiles were pointed afterwards to give the appearance of brickwork. It was popularly believed that they were used to avoid a brick tax, introduced by Pitt the Younger in 1784 (approximately 3 shillings a thousand). In *Mathematical Tiles* by Maurice Exwood (*Vernacular Architecture*, Vol 12, 1984) the author proves that mathematical tiles were in fact more expensive than most plain tiles and bricks at that time. He quotes London area prices in 1862:

Plain tiles per thousand	£2.4s.0d
Malm stocks (bricks)	£2.7s.0d
Common stocks	£2.2s.0d
Mathematical tiles (red)	£3.0s.0d
Mathematical tiles (white)	£3.10s.0d

In other words, mathematical tiles cost approximately 28% more. They were often used to modernize and improve a property. Any savings made in construction would be by the use of a timber frame. The quality of the tiles was such that they were often regarded as more long-lasting than normal brickwork. Mathematical wall tiling is basically imitative of roofing technology, with the same principle of lapping one tile over another on battens. The only major difference is that the tiles are rebated and so, in lapping each other, give the appearance of brickwork. Framing methods used should be stiffer to allow for the use of pointing which needs a more inflexible structure. The mortar mixes should also be on the weak side for flexibility and should also include lime. Any movement that may occur will then be rectified by the slight dissolution of lime in rainwater to heal hairline cracking.

The only other major use of tiles that are dry-fixed and not bedded, is in the pegging or nailing of flat tiles to a wall surface. The most successful method seems to be to fix on the diagonal, leaving a narrow gap between the tiles, which is then mortared over in a half-round section, resembling rounded bars. This type of detail appears in Japan and also in Northern Europe, for instance, in Belgium.

Victorian use of terra cotta owed much more to its successful imitation of stone components, as a 'kit' alternative. The price was approximately half that of using stone (Rivington 1901). At the turn of the century fired clay products became even more popular as an expression of decoration.

Terra cotta is fired but unglazed ware which has a significant proportion of iron oxides. The clays are generally earthenware with a significant addition of ground glass, sand and ground substandard pottery (recycled) which helps to minimize distortion in firing. The body colour of the clay will vary according to how much iron oxide is present. The colour also depends on the degree of firing.

The best terra cotta clay products have an even body composition made from carefully selected and well-graded material. Some terra cotta clays have a body of coarse clay and a face of finer clay. The whole tile is backed with another layer of finer clay to act as a compensating layer and to prevent warping. Sometimes the layers of finer clay were too thin and could spall away from the main body of the tile from differential movement. This method of making tiles is similar to 'encaustic tiles' where the fine baking and facing layers were pigmented and then pressed, the pressed part then filled with coloured slip for a contrasting colour. Burnt tiles were then glazed by dipping into a mixture of powdered glass and water, and re-heated. It was noticed that the more successful and durable coatings were formed when tiles were glazed whilst still hot from the first 'biscuit-fired' process. Some terra cotta has a 'vitrified skin', possibly because smaller particles on the outermost layer have a greater ability to sinter successfully (Fig. 5.1).

Terra cotta has been used to great effect to make highly individual components not only in large

Figure 5.2 Natural History Museum, South Kensington. Cleaning has made these cill details less resilient to water penetration, and subsequent efflorescence and some discolouration is now apparent. (Photograph taken in 1984.)

buildings such as the Natural History Museum, Kensington (Fig. 5.2) but also later on smaller scale on special constructions like the Michelin Building, Chelsea (Fig. 5.3). Built in 1910 and restored in 1987 it is an outstanding example of an Art Deco building. The architect, M Françoise Epinasse of Clermont Ferrand, used a Hennebique ferro-concrete system. The tiles, made by Gilardoni Fils et Cie of Paris, are unique and depict individual racing scenes, with tile details based on automotive parts. The fixing of these tiles is worthy of comment. The complex shapes are three dimensional and were filled with mortar and coursed as blocks, with more complex and weighty pieces restrained by additional metal cramps. These cramps can corrode if exposed to water and air, and any re-furbishment should allow for their replacement.

The next revival in the use of tiles for wall claddings occurred in the 1920s and 30s. They were known as *faience* (a French word for earthenware) although they consisted chiefly of flat earthenware tiles, as used for the Odeon cinemas. The term became synonymous with tiles with a more highly coloured and more obvious glaze.

In the late twentieth century we now prefer plain colour applications as a skin to building. This started with the starkly simple aesthetic of the early modern movement, obsessed with the sterile facade of technology and science. This formed a sharp break with the decorative tradition. The current version of

Figure 5.1 Terra cotta cladding of an office building, Surbiton. Harsh cleaning has removed the vitrified skin from these pilaster details. Water is now more easily absorbed through the face, taking salts and impurities from the adjacent mortar and fixings and encouraging severe discolouration. Cleaning should use only the mildest of cleaning agents and prolonged washing with water, with no abrasives.

Figure 5.3   Michelin Building, Chelsea, London (L.G. Mouchel, 1911) prior to restoration by Conran Roche and YRM. Built using the Hennebique ferro-concrete system. The terra cotta tiles were in good condition when these photographs were taken, although where the vitrified skin had been eroded (by whatever means) there was significant discolouration. (Photographs taken in 1984.)

that aesthetic is the concept of the 'grid', which is in fact very limiting. For a grid to be successful on a building facade it means that the dimensional co-ordination of the building should, ideally, be designed from the outside in and the internal finishes set out accordingly.

One of the leading exponents of this movement, Walter Gropius, wrote: 'Smooth and sensible functioning of daily life is not an end in itself, it merely constitutes the condition for achieving a maximum of personal freedom and independence. Hence, the standardisation of the practical processes of life does not mean new enslavement and mechanisation of the individual, but rather frees life from necessary ballast in order to let it develop all the more richly and unencumbered.'[1]

The use of tiles on the outside of buildings is shown in Figs 5.4, 5.5 and 5.6.

BS 6431:1983 is a comprehensive standard dealing with all aspects of external and internal tiling. There are great problems in tiling which relate not so much to the tiled component as to the supporting building fabric. In effect, the tile is the final component of a composite system of construction and is usually an integral part of the fabric (Fig. 5.7). The specification for tiling cannot be thought of in isolation, and although the tile is fixed last specifications must be developed at an early stage so that repercussions of shrinkage of backing materials, e.g. cements and concrete frames, are fully understood and catered for. Another factor to influence early decision making is the long delivery period of some tiles. Often the best tile is unavailable due to delay in supply and a compromise is selected which is entirely unsatisfactory.

The Italian Ministry for the Environment are currently drafting studies for proposed eco-labelling of ceramic tiles for the EC. The production of tiles is highly energy intensive and it is hoped that legislation

Figure 5.5 External tiling detail. Failure of these tiles has revealed that only a small area gave proper bonding and adhesion.

Figure 5.4 External tiling detail. History Faculty, Cambridge (James Stirling, 1964–7). Water penetration through a horizontally tiled surface and a critical edge junction encourages the leaching out of calcium hydroxide from concrete and cementitious material. This combines with carbon dioxide in the atmosphere to form insoluble calcium carbonate. The projecting window detail acts as a splash zone and cannot be cleaned.

will limit emissions of carbon dioxide and other pollutants and promote greater energy efficiency in the whole process of manufacture, particularly firing. The increase of contaminated water into the river systems, will also be limited and water consumption in washing processes will be much reduced.

Categories of tiles

The European Committee for Standardization (CEN) have agreed a classification for tiles based on water absorption, which is related to their porosity and method of manufacture.

All tiles submitted for inclusion under this standard have to satisfy standards for dimension and surface quality which include the actual dimensions of length,

Figure 5.6 (right) Vulnerable edge detailing in tiles, Cité de Refuge (Le Corbusier, 1933). The junction of several different materials, shapes and surfaces is a potential site for failure.

Figure 5.7 St Thomas' Hospital, London (YRM). Careful detailing showing how tiling is thought of as a composite with the structural fabric. Tiles are clad over individual structural elements with clear separation between. There are no horizontal exposed surfaces and cills weatherproof the tops of in-fill panels. (Photograph taken in 1984.)

breadth and width, rectangularity, surface flatness and quality as defined by the test methods in EN 98, the relevant standard. In every case thicknesses of tiles are specified by the manufacturer, but in order to conform to the tests undertaken for degree of warp or physical strength, it should always be possible to produce a tile of adequate thickness. As a result, there is no clear guidance as to likely thickness from a particular tile size given in the British Standards. The thickness can only be determined after a manufacturer is chosen.

Every category of tile will have a recognizable grouping of types which relate to their quality and ultimate performance. Every group will also have its own modular and non-modular preferred sizes. Table 5.1 shows the classification of ceramic tiles.

Group AI: extruded ceramic tiles with low water absorption ($E < 3\%$)

The extruded tiles may be single and subsequently pressed into shape, or double and subsequently split in two after firing. This category is the highest quality and suitable for all uses, internal or external, in all climatic conditions. These tiles also have the greatest resistance to chemical attack. Certain quarry tiles can come into this category if their composition and manufacture is of a high enough standard.

Group AIIa: extruded ceramic tiles with intermediate water absorption ($3\% < E < 6\%$)

These extruded tiles are split tiles or quarry tiles which may be subsequently pressed. They are suitable for internal and external use but specifiers should satisfy themselves as to the degree of exposure that they can sustain.

Group AIIb: extruded ceramic tiles with intermediate water absorption ($6\% < E < 10\%$)

These extruded tiles are split tiles and quarry tiles. The quarry tiles may initially be formed by extrusion but can be die-pressed into their final shape at a later stage. They are suitable for internal and external use, but are usually used for flooring. This standard is in two parts: Part 2 relates to *terra cuite* produced in France and Belgium, *cotto* in Italy and *baldosin catalan* in Spain. These are slightly softer and less resistant to constant wear.

Group AIII: extruded ceramic tiles with high water absorption ($E > 10\%$)

These extruded tiles are split tiles and quarry tiles. Quarry tiles in this category may also have a later shaping by die-pressing after extrusion. They should not be used where there is a risk of frost.

Table 5.1 Ceramic tile classification

| Shaping | Water absorption | | | |
	Group I $E \leq 3\%$	Group IIa $3\% < E \leq 6\%$	Group IIb $6\% < E \leq 10\%$	Group III $E > 10\%$
A	Group AI EN 121	Group AIIa EN 186	Group AIIb EN 187	Group AIII EN 188
B	Group BI EN 176	Group BIIa EN 177	Group BIIb EN 178	Group BIII EN 159
C	Group CI ...	Group CIIa ...	Group CIIb ...	Group CIII

Source: Based on Table 2 EN 87:1991

All *A* category tiles will satisfy standards whether they are glazed (fully or partly), with glossy or matt finishes, or *unglazed*.

Group BI: dust-pressed ceramic tiles with low water absorption ($E < 3\%$)

These are pressed tiles of high quality. They can be classed as 'fully vitrified' if their water absorption is less than 0.5%. They can also include mosaic tiles and tiles fitted with spacer lugs. They are suitable for interior or exterior use on walls and floors.

Group BIIa: dust-pressed ceramic tiles with intermediate water absorption ($3\% < E < 6\%$)

These tiles are suitable for interior and exterior use on walls and floors.

Group BIIb: dust-pressed ceramic tiles with intermediate water absorption ($6\% < E < 10\%$)

These tiles are suitable for interior and exterior use on walls and floors.

Group BIII: dust-pressed ceramic tiles with high water absorption ($E > 10\%$)

Although used for walls and floors, they are not advised for use where there could be considerable floor loadings or in areas where frost is likely.

The standards all relate to tiles that may be glazed, glossy matt or semi-matt, or unglazed, except in the group BIII which do not have an approved unglazed category.

Accessories

A number of specially shaped tiles can be obtained, such as coves, round edges, angle junctions (internal and external), beads channels, water outlets, etc. Some of the most complex tiles are to be found in specialist

Table 5.2 General definition of dimensions of tiles

Dimensions	Symbol	Modular	Non-modular
Co-ordinating dimension	C	$W + J$	$N_2 + J$ or $W + J$
Nominal dimension	N_1	$W + J$	–
	N_2	–	$N_2 \simeq W$
Work dimension	W	W	W
Joint width	J	J	J

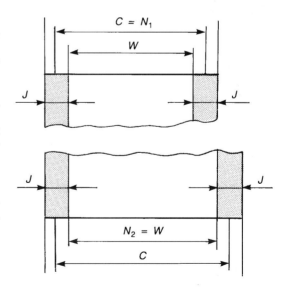

Source: Based on Table 1 EN 87:1991

use for laboratory fittings or in swimming pools, but they can also be used for domestic fittings to comply with austere functionalism. Other domestic fittings include soap dishes which are integral to a tiled wall, dado rails, toothbrush and toilet-roll holders, etc.

Physical properties

Unglazed tiles with a water absorption equal to 6% or more must satisfy standards for water absorption, modulus of rupture, scratch hardness, linear thermal expansion, resistance to thermal shock or crazing, frost resistance and moisture expansion. In addition flooring tiles have to satisfy standards for surface and deep abrasion.

Chemical properties

Tiles must resist staining, household chemicals and swimming-pool water cleanser, as well as alkalis and acids.

Any tiles which come from outside the European Community should be checked to see if they conform with these criteria. Designers should also familiarize themselves with the ASTM standards for these materials if importing from the USA.

Tile dimensions

Tiles today are manufactured in a wide range of sizes. *Modular* tiles are manufactured in metric dimensions, with face sizes in increments of 100 mm or subdivisions of those increments, e.g. measuring 100 × 100 × 5 mm or 200 × 100 × 6.5 or 8 mm. *Non-modular* tiles are tiles which have metric dimensions equivalent to imperial sizes in England, e.g. 152 × 152 mm (6 in × 6 in), or which come in sizes historically produced in other countries but which do not fit modular dimensioning. In working out tolerances and joint widths it is more critical to understand how the tile was made than its overall sizing. Pressed tiles, made from ground particles of clay compressed into a mould are dimensionally more stable than extruded tiles, where green clay is drawn through a mould giving characteristic directional dovetailed ribbing on the back of the tile. The action of drawing the clay through the mould with the consequent directionality of deformation will give a more pronounced shrinkage and movement in that direction. More details about the dimensions of tiles are given in Table 5.2.

In all cases the thickness of the tile is specified by the manufacturer. It depends on the tile size, and is the most economic section that complies with all the physical properties required by BS 6431. Sizes for split and quarry tiles are given in Table 5.3.

Movement and substrates

To minimize any movement in tiling, backgrounds should be completely stable. This is a most difficult criterion to achieve. Backgrounds should be structurally as well as chemically stable and not subject to deflection in use. The whole building fabric should be fully hardened, not just the immediate backing material (the render), and this implies allowing for the hardening of all concrete structural components in the building. Consideration should be given to the natural shrinkage of all cementitious materials, including blockwork, and, where possible, walling components should be chosen to minimize future change or movement. Framed construction systems are usually unsuitable unless proper precautions are taken. Even on a domestic scale it is preferable to tile on renders which are re-inforced with a metal mesh which is made from wires welded together, and are either of stainless steel (austenitic) or galvanized to BS 729. Expanded metal is not as suitable due to the possibility of deformation in the direction of the original expansion. There should be a distinction between using metal mesh as a key and using it as a stabilizing frame to prevent further movement. As this is the type of re-inforcement commonly fixed on stud partitions (whether metal or timber) it can significantly affect the degree of movement. Tiles fixed on frames should have expansion joints where framing systems make connections to other frames or adjacent solid-element components.

Table 5.4 gives a summary of data on backgrounds and various materials for fixing tiles.

Joints

Guidance on joints for tiling is often vague and sometimes contradictory. On the one hand, one tries to specify a weather-tight construction to protect the building fabric, and to allow the passage of water vapour from internal walls to the outside. However, diffusion of gases is a two-way process and depends on many factors, including differential pressure between either side of a material. A general rule is given in BS 5385 Part 2:1978 to allow for a jointing area equivalent to at least 10% of the tile area. This is based on the unlikely passage of water vapour through a fully glazed or vitrified tile. In reality, the recommended

Table 5.3 Sizes for split and quarry tiles

Co-ordinating size *C* (cm)	Work size *W*	Thickness *d*
Modular preferred sizes for split tiles	According to the manufacturer	The thickness shall be specified
M10 × 10 M25 × 25	The manufacturer shall choose the work	by the manufacturer
M15 × 15 M30 × 7.5	size in order to allow a nominal joint	
M20 × 5 M30 × 10	width of between 5 and 10 mm	
M20 × 10 M30 × 15		
M20 × 20 M30 × 30		
M25 × 6.25 M40 × 20		
M25 × 12.6		
Non-modular sizes for split tiles	The manufacturer will choose the work	The thickness shall be specified
20 × 20 24 × 7.3	size in such a way that the difference	by the manufacturer
21.7 × 10.5 24 × 11.5	between the work size and the nominal	
21.9 × 6.6 30 × 30	size is not more than ±3 mm	
22 × 11 40 × 20		
Modular preferred sizes for quarry tiles	According to the manufacturer	The thickness shall be specified
M10 × 10 M20 × 20	The manufacturer shall choose the work	by the manufacturer
M15 × 15 M25 × 12.5	size in order to allow a nominal joint	
M20 × 5 M25 × 25	width of between 3 and 11 mm	
M20 × 10 M30 × 15		
Non-modular sizes for quarry tiles		The thickness shall be specified
10 × 10 20 × 10	The manufacturer will choose the work	by the manufacturer
13 × 13 20 × 20	size in such a way that the difference	
14 × 14 20.3 × 20.3	between the work size and the nominal	
15 × 15 22.9 × 22.9	size is not more than ±3 mm	
15.2 × 7.6 26 × 13		
15.2 × 15.2 28 × 14		
18 × 18 30 × 30		

Source: Based on Tables 1–4 EN 187-2:1991

joints for extruded tiles of 10 mm will easily exceed this minimum area. A joint of 10 mm is often specified as it does allow for good quality pointing although it may be unnecessarily thick.

In BS 5385 Part 2 Section 16.2.3 the advice given is:

'Glazed tiles are impermeable at their surface and vitrified tiles, even when not glazed, have a negligible rate of transfer of water and water vapour. Consequently the size and nature of the joints has to allow for variation in tile size, protection from outside water penetration, and passage of any water vapour from the wall to the outer atmosphere.'

Unfortunately these factors are not quantified and no mathematical relationships which would allow the optimum joint width to be calculated are given. The performance thus depends on the tiler's experience and, unless a specifier has proven information, manufacturers' recommendations should be followed.

Tile composition

All ceramic tiles are made from clay and other minerals usually close to the source of the clays.

Hence kilns and the clay manufacturing industries have historical locations, in order to economize on the transport of raw materials. Clays vary in their texture, mineral composition and colour. Clays fall into groupings: earthenware or stoneware. The earthenware clays vary from reds to browns. There is a classification of 'white' earthenware which is actually creamy in colour. Stoneware is normally fired at higher temperatures and becomes vitrified to some extent, producing a stronger, more durable and less porous finished product. Stoneware clays vary from light greys to light browns ('taupe').

Grading and forming

The raw clay is usually sieved, sorted and ground to obtain a known range of particle sizes that will produce a homogeneous mixture with a particular water content. In a quarry where raw clay is mined there will be considerable variation from the top to the bottom of the clay layers, in terms of particle sizing and general consistency. Clay is usually scraped vertically from top to bottom of the open mine to

Table 5.4 Backgrounds: summary of data and various materials for tile fixing

Background material	Details	Drying shrinkage movement	Surface character	Materials for direct fixing of tiles and mosaics*		
				Cementitious adhesives	Organic adhesives	Cement:sand mortar†
Concrete: *in situ* or precast	Dense aggregate	May vary from low to moderate	Low to moderate suction	S	S	S
	Lightweight aggregate: open surface	Moderate to high	Moderate to high suction	S	S	U
	Lightweight aggregate: closed surface	Moderate to high	Moderate suction	S	S	U
	Autoclaved aerated	Moderate to high	Moderate to high suction	S	S	U
	No fines	Low to moderate	Low to moderate suction	C	C	S
Concrete: blocks and bricks	Dense aggregate‡	Low to moderate	Low to moderate suction	S	S	U
	Lightweight aggregate autoclaved open surface‡	Moderate to high	Moderate to high suction	S	S	U
	Autoclaved, closed surface‡	Moderate to high	Moderate suction	S	S	U
Clay: bricks, blocks and tiles	High-density bricks and blocks	Negligible. May expand slightly	Low suction	S	S	U
	Normal bricks and blocks	Negligible. May expand slightly	Moderate or high suction	S	S	U
	Tiles and glazed bricks	Negligible. May expand slightly	Very low suction	C	S	U
Calcium silicate bricks	Hard bricks	Low to high	Moderate suction	S	S	U
	Soft bricks	Low to high	Moderate suction	S	S	U
Natural stone	Hard stone	Negligible	Low suction	S	S	U
	Soft stone	Negligible	Moderate or high suction	U	U	U
Cement:sand rendering	New	Moderate	Moderate suction	S	S	S
	Existing	Negligible	Moderate to high suction	S	S	S
Other surfaces	Fibre cement board. Wood-based panel products	Moderate to high	True and smooth	C	S	U
	Paintwork	Not applicable	Not applicable	U	U	U
	Metal surfaces	Nil	Low suction and poor key	U	S	U

Note: For detailed advice on the preparation of backgrounds refer to BS 5385 Part 2:1991 Section 18.

* S denotes 'suitable', U denotes 'unsuitable' and C denotes 'confirm adhesive's suitability from manufacturer'.

† Cement:sand mortar is seldom used for fixing tiles externally and has been included in this table as an alternative method of fixing mosaics. For fixing tiles in cement:sand mortar see BS 5385 Part 1:1990.

‡ Properties of backgrounds indicate only relative characteristics of the materials.

Source: Based on Table 3 BS 5385 Part 2:1991.

Figure 5.8 Clay ridge tile. These cavities are caused by erosion but show clearly the finer vitrified skin on the surface and the coarser backing material in the body of the tile.

achieve as good a mix as possible before grading. The material is then formed into regular shapes by pressing or casting into moulds, or is extruded through controlled section profiles and cut to the required length. The products formed are then left to become 'leather-hard' by evaporation of water with subsequent shrinkage. After this the tiles are then fired and the clay particles will sinter until they partly fuse or completely coalesce to give the hard and rigid final product. *Split tiles* are made from extruded forms, and the double tiles are split after firing.

Firing

The first firing can produce a coloured tile if *slip* colours are used. A slip is a fine slurry of clay applied at the leather hard stage to give an integral colour to tiles. Glazing is usually applied on the first-fired *biscuit ware* and a second firing at a higher temperature will fuse the glazed coating on to the body of the biscuit ware. The glaze applied is a solution containing a mixture of silica (silicon dioxide), one or several basic oxides to alter the actual colour and alumina (aluminium oxide) to give some viscosity to the solution. Calcium, barium, lead, magnesium, potassium, sodium, zinc are the basic oxides used.

Materials for fixing

Care should be taken in the specification of mortars and all their constituent ingredients. The performance of tiling relies on a very exacting specification. Some of the main causes of failure are due to the use of impure materials or to incompatible specifications for tile and background material. All ingredients for mortar must be clearly referenced to British Standards (e.g. cement to BS 12:1991) for purity. Sand is the most likely source of contamination: salts or other matter can affect the final dimensions of the joint by unpredictable shrinkage.

Backing coats, pointing mortars and the aggregates for these mixes should all be inert to minimize chemical degradation. Fine particles of *opaline silica*, a constituent of chert, flint, shale sandstone or limestone, lead to the formation in the alkali/aggregate reaction of an impure gel with a greater volume than usual. In fixing their tiles, the use of a chromalith mortar which is highly refined and is imported from Luxembourg is recommended. It contains quartz silver sand, limestone, pigment and a waterproofer.

The careful shaping of joints is critical to water run-off (Fig. 5.9). Recessed joints should be avoided, and flush-faced or bucket-handled joints are preferable.

Refer to BS 5385 Part 2:1991 for advice on backgrounds, treatment, backing materials, bedding methods and adhesion. Refer to Table 2 abstracted from BS 5385 for properties of different flexible sealants. The BRE have produced a Defect Action Sheet no 137 (published in November 1989) on the loss of adhesion between tile and substrate. This emphasizes how movement causes failure of adhesion with small scale movements having a disproportionate effect on poorly bonded tiles.

5.2 Concrete finishes

To obtain a good finish, whatever the texture, it is essential to specify quality ingredients with a sound specification for formwork and the placement of

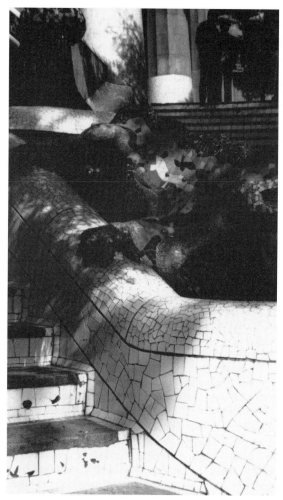

Figure 5.9 Parc Guell (Gaudi). Extraordinary, though technologically sound detailing. Exposed surfaces are curved so deposits and water are flushed away naturally. The ceramic pieces are small and already fractured into energy-efficient shapes.

concrete. Concrete often displays a whole range of blemishes which can be unsightly and which can all be traced back to poor mix design or bad workmanship. Unacceptable appearances usually due to a combination of factors rather than one single factor: mix design, curing, release agents, retarding agents, weathering in the course of setting, excessive vibration or striking formwork too early or too late may all be involved. All of these factors can be controlled but need a good specification and supervision.

For a complete introduction to concrete see *MBS: Materials* Chapter 8. Also refer to *Introduction to Concrete* and *Concrete Practice* (Cement and Concrete Association).

Further reading

The British Cement Association publishes the following:

Visual Concrete: Design and Production William Monks.
The Control of Blemishes in Concrete William Monks
Efflorescence on Concrete D D Higgins
Textured and Profiled Concrete Finishes William Monks
Exposed Aggregate Finishes William Monks
Tooled Concrete Finishes William Monks
Structural Concrete Finishes: A Guide to Selection and Production William Monks

The prevention of major blemishes is an important area, which we will tackle next.

Formwork

Formwork should be stiff enough. It must be inflexible with a uniform face and very little absorbency, as water is needed by the chemical reaction in the mix. The joints should be water-tight and the formwork should be evenly coated with a release agent and free from any material that can stain the mix, e.g. fixings from the formwork. All formwork should be sealed – especially timber which should have lapped joints where possible to avoid excessive water absorption from the concrete.

It is surprising how many designers will specify British Standards for most of the work without realizing that the materials for the formwork may still be unspecified and left to a contractor unless defined precisely by the designer. If the formwork is struck and the finish found to be unsatisfactory, it is very difficult to make good, so care must be taken in the drawings and specification to ensure that the right finish will be achieved. It is worth specifying that *sample panels* should be cast for fairfaced work for approval, before proceeding with the main work.

A variety of materials can be used for formwork as long as they are constructed to be stiff enough to take the weight of wet concrete. Standards of formwork should be high, as the skills required are those of a joiner rather than a carpenter for the best finishes. Timber, metals, plastics, rubber, polystyrene and even hardboards can be used for formwork and very fine textures can be reproduced. Timber is often sandblasted so the grain is emphasized. Finishes should also be designed to take account of weathering. Very smooth surfaces are unsatisfactory in this country. Textured surfaces with a vertical emphasis are better for deliberately channelling water flow. Horizontal features will cause streaking unless proper drips are introduced.

Retarders Chemical retarders are applied on formwork to ensure that, although the body of the concrete has set, there is a slurry on the surface of the concrete that can be removed easily to expose the underlying aggregate. Most of these retarders are sugar-based and are difficult to use. Results are not always predictable and retardation can inhibit in-depth setting of the mix.

Acid etching Fine surface detail can be achieved by etching with a 5% solution of hydrochloric acid. The length of time it is left on the surface will determine the degree of etching. This type of finish is probably more useful for altering the texture of an existing building, as the same degree of control can be reproduced by careful choice of formwork. Surfaces treated in this way need thorough rinsing and adjacent surfaces, especially glass must be protected.

Concrete mix

It is very important that accurate mix proportions are used: a low sand content will create voids and a cement-rich mix will be more likely to craze. Dry or porous aggregates will take up too much of the water content of the mix leading to a less dense mix with a friable and sometimes powdery surface. Too much water will also lower the strength and produce voids. The mix must be consistent and excessive vibration during setting should be avoided. See BS 8110 Part 1 *Code of Practice for Design and Construction*, BS 5328:1991 *Methods for Specifying Concrete Including Ready-mixed Concrete*. Note that new methods of specifying concrete take into account site conditions. See Table 13 of BS 5328 Pt 1 and for more specialized use, Table 6 of BS 5328 Part 2.

Curing

Most problems in curing concrete stem from an inadequate understanding about how concrete hardens. The action of mixing water and cement initiates a *chemical reaction* known as *hydration*. The water is an essential component which has to be mixed in the correct proportions, and never mixed to achieve a degree of workability. As the water is needed for the reaction, steps must be taken to ensure there is no appreciable water loss, so concrete is protected while setting to prevent loss of water from evaporation.

Curing should be even: too rapid or inadequate drying will give differential moisture loss. Weather conditions should be carefully monitored otherwise uneven hardening will take place.

White concrete

For white concrete white cements should be specified, but all the aggregates should also be specified in lighter colours, e.g. light limestones and granites, calcite spar or calcined flint with fine silica sand as the fine aggregate. This does lead to an expensive and possibly uneconomic mix, and demands a greater degree of control to minimize rates of shrinkage that are prevalent with white cements.

Summary of blemishes

Hydration discolouration There are often variations in colour at the joints of formwork or sometimes patchiness, echoing the grain in timber or plywood or as a shadow pattern of aggregates. All of these colour variations are due to *hydration discolouration*. If there is a differential moisture loss from the body of the concrete while it is setting there will be a high-concentration of cement and a marginally lower water content in that specific region. This means a slight change in the type of compounds formed and a greater proportion of tetracalcium aluminoferrite, which has a darker grey colour than normally set Portland cement with the correct water–cement ratio. So areas where moisture is lost will always be a slighter darker colour. If different veneers or timber are used and not properly sealed there will be different rates of absorbency which will also give a different colour to the cement when set. Aggregates close to the surface of the concrete that absorb moisture from the mix will give a colour difference and a shadow effect for the same reason. Sometimes this is called 'segregation discolouration'.

Lime bloom In the hydration of the cement tri-calcium silicate and di-calcium silicate can liberate free lime (calcium hydroxide) which can be leached out of the concrete by rainwater or moisture and react with carbon dioxide in the atmosphere to form the more insoluble calcium carbonate at the surface. This deposit can be removed with a 5% solution of hydrochloric acid but if the concrete is continually saturated the fault may persist and may reflect a detailing problem.

Both the above faults are present in the building shown in Fig. 5.10.

Figure 5.10 Unite d'habitation, Firminy Vert (Le Corbusier, 1967). This concrete wall is on top of the building on a high roof terrace and shows a number of faults. Age enhances the definition of formwork joints (hydration discolouration). Mirror cracking repeats the structural form on the other side of the wall giving pathways for the external deposition of lime bloom (deposits of insoluble calcium carbonate).

Crazing   Crazing is a pattern of fine shrinkage fractures on a cement-rich skin to the concrete. It is usually due to using too smooth a formwork finish. More absorbent formwork surfaces can raise fine aggregates to the surface but these have crack-stopping capabilities and prevent large areas of cement-rich skin from being under stress. It can also be due to an excess of water in the mix. Rather than repairing the surface, fine sandblasting or acid etching to form a slight texture may be more successful.

Blow holes   These defects can be caused by using surface formwork which has no absorbency. Too little vibration is also a cause and leaves air pockets trapped against the side of the formwork.

Scabbing   On striking the formwork a very rough face to the concrete with tearing of the material indicates a poor application or performance of release agents.

Pyrites   Pyrites are iron-rich aggregates (containing iron sulphide crystals). They can be the cause of streaking rust-brown stains down the face of the concrete. They can also react expansively with free lime in the concrete forming iron hydroxide which has a greater volume. The increased volume of material can cause cracking. It is difficult to detect these aggregates prior to staining or reaction and the only remedy is to pick them out. If the area is localized and the source of the staining seems to be a soft grey material, then this is confirmation of a reactive aggregate. Staining of this nature can be confused with staining from projecting small wire reinforcement ties. Prevention should be by checking the purity of the aggregates.

Dusting   If the concrete is not properly dampened down, excessive water loss can cause rapid hardening and a friable surface.

Major deterioration and repair

The first step in the repair of faulty reinforced concrete, is establishing a diagnosis. The first indications of failure are usually cracking and/or spalling of the concrete. If cracking is parallel with the reinforcement, it is indicative of expansive corrosion of reinforcing steel. The depth of cover can be established using a cover meter and the reinforcement plotted. If the cracking patterns do not follow the reinforcement then a separate structural investigation needs to be carried out. As a brittle ceramic, it is likely that cracking will occur in directions that are perpendicular to the tensile stress applied.

The concrete cover should be carefully removed to look at the nature of the corrosion. If the corrosion of the reinforcement is even, i.e. to the same depth along the length of the steel, this suggests carbonation, whereas corrosion showing severe pitting indicates chloride attack.

Carbonation   Carbonation is a term used to describe the progressive change of calcium oxide in the original concrete mix into calcium carbonate. The pH of the porewater changes from a value between 12.5 and 13.6 (alkaline) to a value of 3 (acid). The reinforcing steel, which in a *passive* alkaline environment is not susceptible to corrosion then becomes liable to corrosion. The relevant reactions are as follows:

$$CO_2 + H_2O \rightarrow 2H^+_{(aq)} + CO^{2-}_{3(aq)} \ldots \text{Dissolution of}$$
atmospheric carbon dioxide to form carbonic acid
$$CaO + H_2O \rightarrow Ca^{2+}_{(aq)} + 2OH^-_{(aq)} \ldots \text{Reaction of}$$
calcium oxide with water to form a basic solution of caclium hydroxide

$Ca^{2-} + CO_3^{2-} \rightarrow CaCO_3$ ... Precipitation of calcium carbonate

$H_2O \rightarrow H^+_{(aq)} + OH^-_{(aq)}$ ... Water to hydrogen and hydroxyl ions.

Since carbon dioxide is present in air and the current contains calcium oxide the reaction will proceed. The critical point to establish is the rate at which carbon dioxide travels through the concrete. This is the *rate of diffusivity* and it should be used to define the depth of cover needed. The recommended depth of cover changed from 25 mm in 1948 to 50 mm in 1972.

To determine the extent of carbonation that has already taken place in a sample of concrete surrounding corroded reinforcement, a technique using a solution of phenolpthalen in diluted ethyl alcohol is described in BRE information paper IP 6/81. If sprayed on a freshly fractured surface, it should change to a healthy pink if the concrete is still alkaline (pH > 9). It will stay colourless if the concrete has a pH value of less than 9. If the test is alkaline then samples of

Figure 5.11 Carbonation and its effects. This corrosion of this reinforcing rod is the result of carbon dioxide diffusing through the concrete matrix and changing the pH of the pore water in concrete from an alkaline pH of about 9 to an acid value of 3. In this new acid environment the reinforcing steel has a greater potential to corrode, with subsequent expansion, fracturing of adjacent concrete and eventual spalling. It results from inadequate concrete cover to steel. At this stage the steel will have to be cut out and new parts welded in, with priming and keying to take a matched new mix. Testing for carbonation can be carried out on site with phenolphthalein which will show healthy alkaline concrete to be pink. Cracking due to carbonation takes place parallel to the reinforcement.

concrete should be tested for chloride content. (BRE information sheets IS 12/77 and 13/77 and BS 1881). Figure 5.11 shows an example of a reinforcing rod corroded by carbonation.

An example will illustrate the scale of the problem. Extensive repairs were carried out at Teeside Polytechnic after deterioration of the concrete frame was confirmed as being due to carbonation. The building was only 20 years old and carbonation was found to have occurred at least to a depth of 8 mm and in some cases up to a depth of 40 mm. Concrete cover varied from 10 to 40 mm for main steel, and nil to 12 mm for secondary steels. The general strategy for repair was to expose all the reinforcement where corrosion was taking place, clean by grit blasting and then make good. This was achieved by initially protecting the reinforcement with a cementitious grout with a corrosion inhibitor, giving a new alkaline environment to the steel, then applying a slurry with a bonding agent, before finishing with a repair mortar. The whole frame was finished in a coating which was impervious to the passage of carbon dioxide but which allowed the diffusion of water vapour. The coating is anticipated to last for 10–15 years before reapplication is required. (Source: *A new lease of life for a reinforced concrete building* by Charles Morris. *Construction Repairs and Maintenance*, July 1986.)

Chloride attack Concrete may be alkaline but corrosion can still occur if there are free chloride ions. Chloride ions can form stable compounds with calcium aluminate in the mix, but there is a limit to the amount of chloride ions that can be accommodated in this way. Excess ions are free to travel and transform pore water into an electrolyte, which dissolves metal ions. Chlorides are usually introduced into mixes as contaminants in aggregates (sea-dredged) or as an accelerator in the form of calcium chloride. Chloride ions may also come from proximity to marine environments or from the de-icing of roads in winter. Calcium chloride was banned as an accelerator in 1977 but remedial work has to tackle the repercussions of earlier usage.

BRE digest number 264 and BS 12 (in relation to the cement content for dense aggregate concretes) specify the following categories of chloride content:

Low chloride ion content: up to 0.4% by weight of cement
Medium chloride ion content: from 0.4 to 1%
High chloride ion content: over 1%

Corrosion of the reinforcement is likely if the ion content is over 1%.

Refer to EN 196 Pt 21 for determination of the chloride, carbon dioxide and alkali content of cement. Part 27 describes a method for determining CO_2 content using phosphoric acid to decompose the carbonate present. Alternatively, sulphuric acid can be used, releasing CO_2 which can be measured after absorption by sodium hydroxide. All site testing should use safe preparations.

Alkali aggregate reactions Concrete has the capacity for further chemical change by the action of alkaline pore fluids in the concrete on siliceous aggregates. The end product is a calcium silicate gel which absorbs water and swells, causing cracking. This is not a common reaction and not many cases have been reported in the United Kingdom. The known cases are confined to the South-West and the Midlands in England. Several common factors including high alkalinity, high cement content and exposure to water are present.

Core samples from suspected cases may be coated with calcium silicate gel. Alternatively sections may be taken for microscopic examination and cracked aggregate particles with their cracks filled with gel can be clearly seen.

There is little evidence in repair work to structures damaged by this reaction, but, if diagnosed, efforts should be made to improve the water resistance and water penetration of the concrete to avoid absorption by the newly formed gels.

The best prevention is selection of the cement aggregate and use of low-alkali cements and cement-replacement materials. The most reactive aggregate is opaline silica which is not common in the UK. Other reactive aggregates are likely to be found in sand and gravel off the South Coast, the Bristol Channel and the Thames Estuary and in parts just inland from these areas. They include microcrystalline and crypto-crystalline silica, chalcedony and some quartz. *Alkali-aggregate reactions in concrete BRE Digest 330*, March 1988.

Repair and guidance

BRE Digests 263, 264 and 265 (*The durability of steel in concrete*, Parts 1, 2 and 3) outline the two major causes of corrosion to reinforcement (carbonation and chloride attack) and recommend strategies for repair. The Concrete Society Technical Report No 26 (October 1984) elaborates on these strategies.

If carbonation has been diagnosed, then repairs are possible and do increase the longevity of the building. The following points apply:

1. Cut back all concrete around corroded material to a depth of 12 mm behind the steel and 50 mm beyond the corroded area, after taking the load off the element being repaired by jacking;
2. Clean all corroded deposits from reinforcement and weld in new steels if the degree of corrosion reduces primary reinforcing members by 10% of their diameter;
3. Restore a passive environment to the steel either by using barriers to aggressive agents or by raising the alkalinity of the immediate environment to the steel;
4. Incorporate a bonding agent to prevent subsequent spalling;
5. Apply the new repair material to match the strength of the existing mix. It must be chloride free, and be installed in thin layers. It is generally recommended to use a Portland cement modified by a polymer latex for ease of handling and reliable performance. Epoxy-resin mortars and grouts as well as poly-ester-resin mortars and grouts are more difficult to use. Epoxy resins have lower shrinkage rates than polyester resins, but are more expensive.
6. For information on surface coatings, BRE Digest 265 gives a comprehensive table of types and their performance. When carrying out repair work and relying on a coating finish, supervising officers should note that paints can only bridge over crack widths up to 0.2 mm maximum.

Dense concrete mixtures help protect steel, but we should be limiting the movement of ions by controlling the *permeability* of concrete. Coatings offer a very different order of permeability due to the nature of their close chemical bonding. A polymer coating of 100 microns in thickness can be from 20 to 200 times more effective in terms of resistance to movement of CO_2 than a 50 mm thickness of concrete. With regard to the protection of concrete from water, paint films will not absorb water although concrete will. The difference in the passage of water *vapour* is less dramatic. Consequently paint films do offer a high degree of protection from externally absorbed water which may contain chlorides and gases in solution. *Permeability and protection of reinforced concrete* by C.D. Lawrence. Cement and Concrete Association reprint 6/86, from symposium *Concrete structures – the need for protection*, Wakefield 1984.

Panels

Concrete panels are more vulnerable to deterioration from carbonation as they are usually exposed on both sides to the atmosphere.

Figure 5.12 Villa La Roche-Jeanneret (Le Corbusier, 1925). Reinforced concrete pylons with cement-rendered breeze block walls. A well-maintained facade, but note that the windows have clear cill details and horizontal surfaces are weathered. (Photograph taken in 1986.)

Moulds for panels are expensive and if possible, the design should be standardized to reduce the number of different panel types. Steel moulds are the longest lasting but also the most expensive to make. GRP has a longer life than timber but less than steel. Handling on site is an important consideration and panels should not be made heavier than 7 tonnes, otherwise there will be problems in delivery and cranage.

The thickness of concrete panels could be slimmer than that adopted at present for equal structural strength. Thickness is governed by cover to steel. Glass-reinforced cement panels have become an acceptable alternative for cladding, with the advantage of being far lighter in weight. Panels can be cast over trays of selected aggregate bedded in sand as an alternative to casting into moulds.

The decision to use panels requires the designer to consider concrete joinery where not only the cast finish, but also the method of fixing and setting out, and the design of weatherproof junctions all need specifying. It is advisable to work with a cladding manufacturer from an early stage.

Finishing treatments on hardened concrete

Abrasive blasting The surface finish and degree of texture obtained depend on the size and type of particle used in a stream of air or water. This can be sand, grit or shot. It is now common practice to use water in preference to air, as it is safer for operatives.

Tooling Mechanically operated tools are used for bush hammering but have to be handled carefully to avoid shattering the aggregates. This kind of mechanical treatment should not be carried out until the concrete has hardened for at least three weeks.

Grinding and polishing Finer work can be achieved by point tooling, or by grinding and polishing two to three days after casting. The greater the labour needed, the more expensive the finish, and this provides an additional argument for making sure that the formwork details are precise enough to minimize time-consuming hand-finishing.

5.3 Renders

There are great similarities between plastering and rendering, except that renders are usually used for external work and their final mix depends on prevailing weather and microclimatic conditions.

The cementitious materials set by hydration and so water is needed for the chemical reaction. This is why the materials in their powdered form should be properly stored on site away from moisture, and kept under cover and off the ground. They are also particle composites with aggregates acting to stop cracks.

Renders have become less successful recently in many instances, despite a long history of use in traditional building. This is due to a complete change in the type of materials used to achieve a homogeneous but relatively thin coating, and also to a less rigorous approach to creating exactly the right conditions for applying the render. We now expect a higher standard of finish, and the optimization of physical dimensions. The surface area to be covered may be so large that problems of movement, whether caused by stress or by thermal response and flexure, must be catered for in a way which will show on a building facade, in the setting out of expansion and contraction joints. These lines will be very apparent and are strongly resisted by architects to the detriment of the final finish. There is also a preference for smooth finishes which are not as reliable in terms of overall performance as the more textured finishes.

Render is also expected to be successful on a variety of substrates, from concrete to brick. It is applied to block and even over framing systems. Render is often erroneously thought of as a material in its own right which is simply applied over all these different types of construction. Every time the substrate changes, the nature of the render coating used should also change. Render and substrate must be thought of as a *unified composite material* compatible in terms of its likely physical behaviour under stress. After all, both skins must behave as one (Fig. 5.13).

The Code of Practice for render is BS 5262:1991 *Code for external rendered finishes*. The latest issue of the standard takes account of failures in renders and gives recommendations for polymer-modified coatings. An important addition is the technique of rendering over external insulation, a technique widely used on the Continent. External insulation is the most effective position for energy efficiency, but the render must contain additives to give greater flexural strength. Special mixes may include polymers, fibres and mixes for application over mesh or lathing. These are sold as pre-mixed render coatings. The use of additives helps achieve higher bond strengths, better water resistance and general durability. The Code does exclude special advice on renders for tanking and for liquid-retaining structures. The Cement and Concrete Association publish a booklet on *External Rendering* (7th edition) which gives practical advice well-illustrated with photographs.

Definitions of *bond, suction* and *adhesion* are the same as for plaster. All materials should be pure and conform to British Standards, as follows:

> Cements to Portland cement BS 12:1991
> Specification for Portland blast furnace cements BS 146:1991
> High alumina cement BS 915 Part 2:1983
> Sulphate-resisting Portland cement BS 4027:1991
> Masonry cement BS 5224:1976 (Although based on Portland cement this cement is more plastic through air-entrainment and better at retaining water).
> Building lime to BS 890:1972
> Building sands from natural sources to BS 1199:1976
> Ready mixed material to BS 4721:1981
> Aggregates from natural sources to BS 882:1992
> *Note*: new terminology now replaces *fine aggregate* with *sand*, meaning natural, uncrushed, partially crushed and crushed rock. *Fines* replaces silt, clay and fine dust, or any material passing through a 75 micron sieve. *Coarse aggregate* refers to materials passing through a 14 mm sieve.

Renders contain other components, such as waterproofers, workability agents, bonding agents, and water-retaining agents.

Specification

The specification depends on the substrate and the degree of exposure of the building. This ranges from sheltered, through to moderate or severe exposure. As the render is external, the specification must cope with all seasonal variations of temperature and moisture to avoid deterioration. Over-rigid mixes are not to be advocated as they can crack under stress and water will

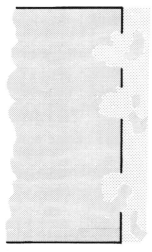

Figure 5.13   Render and substrate as a composite. The outer skin will deform as a result of inner stress.

penetrate through hair-line fractures at a fast rate due to capillary action. Open, porous textures and rough textures seem to weather the best. There is also evidence to suggest that a larger scale of porosity is not so vulnerable to freezing conditions. (See section 4.5 on frost and freezing mechanisms.)

The design of buildings has sometimes sought for a crispness and precision 'reminiscent' of the Modern movement without proper research. Some buildings by Le Corbusier and Frank Lloyd Wright have details that can only be appreciated if the working drawings have been inspected or the buildings visited. Parapet details were sometimes curved to give adequate run-off of water. Render details by de Koninck still incorporated methods for water run-off around openings. The detailing of all *edges* should be a paramount consideration when render techniques are used, and details should be designed at an early stage for:

Parapets and copings

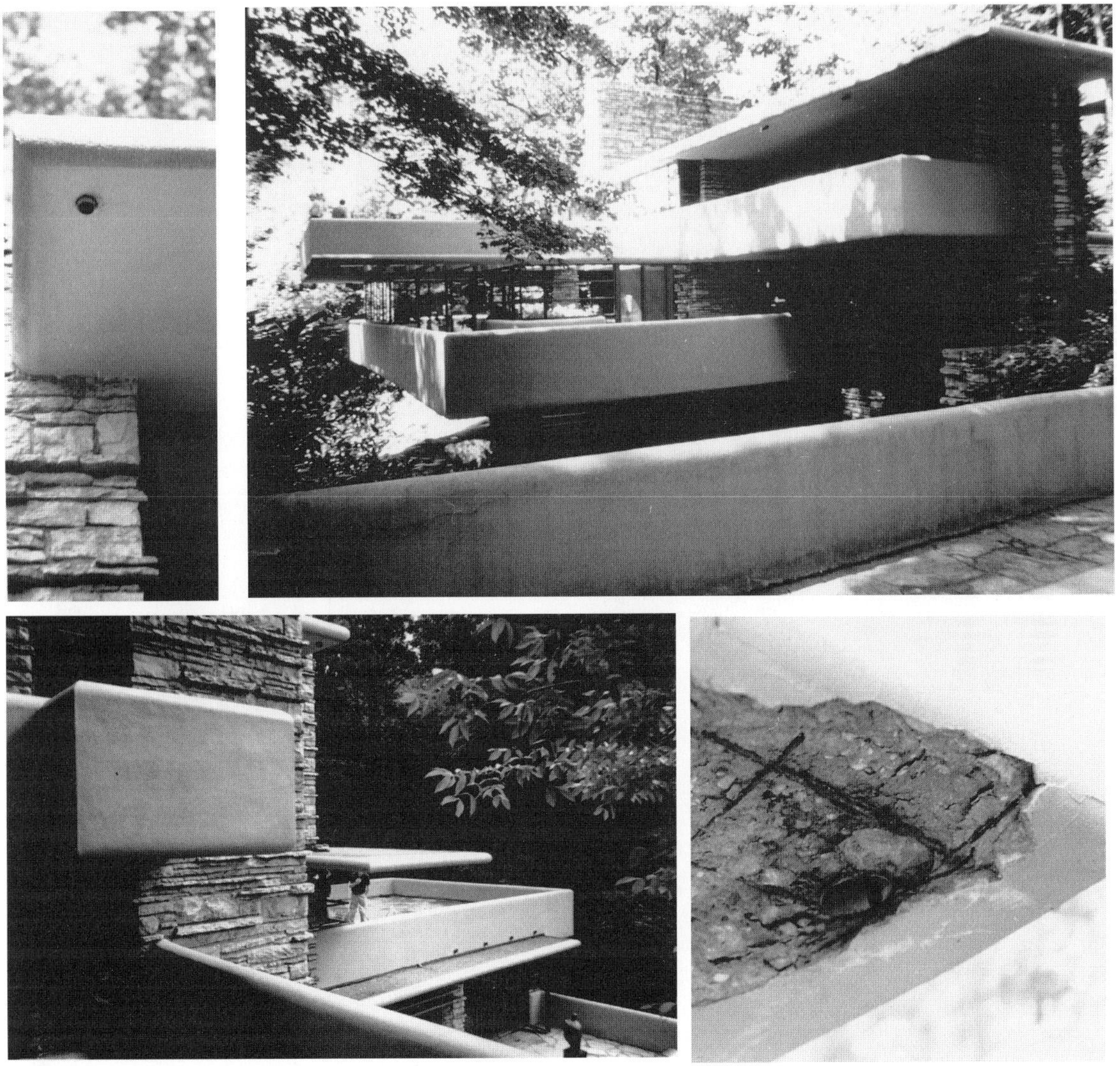

Figure 5.14 Edgar J. Kaufmann House, Bear Run, Pennsylvania (Frank Lloyd Wright, 1936). All the rendered surfaces are curved (*adobe* fashion) helping water to flow off the building. With an absence of drips the water can follow the curved surfaces. *Top left*: Drainage has to be carefully detailed as deterioration can occur if the outlet does not throw the water clear. *Bottom right*: Water intrusion and rusting of the nearby and inadequately clad mesh reinforcement has caused cracking and spalling. Small-scale stalactites of calcium carbonate have formed as free lime is exposed to carbon dioxide in the atmosphere.

Figure 5.15   Sanatorium 'Zonnestraal', Hilversum, Holland (J. Duicker, B. Bijvoet and J.G. Wiebenga (Engineer) 1926–31). This classic Modern movement building shows the tenuous early detailing of mesh-reinforced walling in-fill to a reinforced concrete frame. *Top left* and *middle*: The disused wing of the hospital and progressive deterioration through carbonation and penetration of the building fabric by water. *Top right*: The in-fill panels show the lightweight sandwich of render over steel mesh. Rusting metal bulges and puts stress on these light in-fills. These buildings need good maintenance, as deterioration of their finishes has severe repercussions. *Bottom*: Other parts of the same complex when good maintenance has been provided and the original intention of the buildings is clearly expressed. (Photographs taken in 1985.)

Openings around windows and drips above doors
Details that abut damp-proof courses
Arrises
Details of eaves

The principal aim is to get rid of water and stop it from flowing back into the building. For example of render on modern movement buildings, see Figs 5.14 and 5.15.

Substrates and surface treatments

For general information the BRE have issued a good building guide (GBG) No 18 (June 1994) which looks at relative exposure of renders across the UK, and gives recommended coatings for differing substrates suitable for different degrees of exposure. The information is presented in a pictorial way, is easy to understand and also gives construction details for cills, eaves, around openings and finishing at the base of walls.

The Code of Practice for external renderings is BS 5262:1991. The backgrounds are categorized in a similar way to plasters and this standard should be used as a source for detailed reference, although the following paragraphs give an outline. For all backgrounds initial preparation and restoration of the fabric must be thorough to make sure that bonding and strength is optimized, and poor adhesion or cracking over defective backgrounds is eliminated.

Solid substrates

High density clay, concrete bricks, blocks, dense concrete or low density concrete with a sealed surface   These are all low-porosity surfaces to which it is difficult for render to adhere. In these situations keying surfaces should be provided wherever possible: keyed bricks and blocks, metal lathing, bush hammering or other physical or chemical methods should be used to expose the aggregate or a spatterdash coat applied to provide an initial roughened surface for further coats. Retarders can be used with concrete to delay the setting of the outer face of the cement so that, when the formwork is struck, the unset slurry can be brushed over to provide a mechanical key. If retarders have been used the surface should be well washed afterwards to provide a clean surface.

Bonding treatments are mentioned but not advocated in the British Standard due to the lack of long-term experience of the materials and the vulnerability of some polymers in below freezing conditions.

An example of rendering over reinforced concrete is shown in Figure 5.16.

Moderately strong but porous substrates

Clay, calcium silicate bricks and porous lightweight concrete   These materials have some key and also show good adhesion as the surface topography is sufficiently structured to generate good capillary action.

In materials that are jointed, the joints should be well raked out and all debris removed from the surface, including efflorescence. Old brickwork should be raked out to not less than 13 mm depth. If the mortar is too hard to rake, an alternative treatment is hard-scoring of the brickwork, or the application of metal lathing. Metal lathing may be needed over brickwork which is deteriorating to such an extent that it will not support a render.

Calcium silicate bricks and concrete blocks may be sufficiently porous without any treatment.

Woodwool slabs   Woodwool slabs might be used as permanent formwork and then need a finishing surface.

The joints in this material are usually the greatest problem. They should be reinforced and bridged over with expanded metal, unless it is intended to use metal lathing over the whole surface. Fixing into woodwool is very difficult and clipping will be more successful than nailing even at an angle. For channel-reinforced woodwool slabs, the junction should be covered with 250 mm wide expanded metal. A spatterdash coat is recommended as the first of a three-coat application thrown on with frequent damping in warm weather to assist full hydration.

Metal lathing   Rendering is used over metal lathing on a timber-frame construction or over external insulation.

It is critical that the corrosion resistance of the lathing is adequate. Three-coat work is specified in the standard as being the minimum required on metal lathing to provide a coat adequate to resist weather penetration. Of course, the depth of render required must also be sufficient to provide a rigid matrix that is less susceptible to cracking by preventing bending of the lathing.

The fixing of metal lathing should be adequate for the gauge and type of lathing used. As a general rule, ordinary expanded-metal lathing will need supports at 350 mm centres minimum. Lath with metal ribbing, or welded mesh, is stiffer and can span further. Individual manufacturers' instructions should be followed for centres. Mesh sheets should always properly lap over each other and be tied together at the edges to prevent movement through the render.

Figure 5.16   House and detail for Mr Dotremont, Brussels (L.H. de Koninck 1931–2). The building is in reinforced concrete with rendered walls. Mesh reinforcement is on a 200 mm grid. Formwork is boarded (approximately 100 mm wide) prior to rendering with a rough aggregate finish. This is still in a reasonable condition. (Photograph taken in 1983.)

All metal lathing should be made from austenitic stainless steel, and galvanized components should only be used in sheltered conditions. See the related British Standard 1369:1987 *Steel lathing for internal plastering and external rendering*, BS 1449 Part 2:1983, Stainless and heat resisting steel, and also BS 405:1987 *Specification for uncoated expanded metal carbon steel sheets for general purposes* for expanded metal. Welded wire mesh should be made of squares between 25 mm and 50 mm in size, and from not less than 1.02 mm gauge steel, and should be made of galvanized or stainless steel.

There will be far greater movement of expanded metal over timber and metal studding constructions. Metal lathing in these circumstances is not normally stiff enough to cater for the dimensional movement of timber or the thermal movement of metals, and additional restraint fixings have to be designed. Perhaps a special type of lathing which has a cross braced pattern should be used. If applied to timber construction metal lathing should be fixed over building paper or other material which can act as a waterproof breather membrane.

Mixes for rendering   As a general principle mixes should become weaker towards the outer face of the building. This allows for greater flexure on the outer skin. Finishes which have exposed aggregates and rougher textures are less liable to crack due to the crack-stopping nature of the mix and consequently are more weatherproof. The techniques of application are important and care should be taken to make sure that successive coats harden slowly and do not lose moisture, otherwise a weak coat will result and cracking will be more likely.

See Table 1 of BS 5262:1991 on which Table 5.5 is based. Mix type 1 is extremely strong and can only be used as a first coat on very stable surfaces, e.g. concrete, or as a stiff first coat giving rigidity to expanded metal. The other mixes (2,3,4) are not so hard and are less likely to crack.

Table 2 of BS 5262 gives a complete guide to the degree of exposure and substrate of renders. See BS 5262 for alternative mixes if required. Clause 22 of this standard gives guidance as to the relevance of choosing mix types I to IV. Type I mixes are the strongest and to be avoided due to risk of cracking,

Table 5.5 Mixes suitable for rendering

Mix designation	Mix proportions by volume based on damp sand				
	Cement:lime:sand*	Cement:ready-mixed lime:sand*		Cement:sand* (using plasticizer)	Masonry cement:sand*
		Ready-mixed lime:sand	Cement:ready-mixed material		
I	$1:\frac{1}{4}:3$	1:12	1:3	–	–
II	$1:\frac{1}{2}:4$ to $4\frac{1}{2}$	1:9	$1:4–4\frac{1}{2}$	1:3–4	$1:2\frac{1}{2}–3\frac{1}{2}$
III	1:1:5 to 6	1:6	1:5–6	1:5–6	1:4–5
IV	1:2:8 to 9	$1:4\frac{1}{2}$	1:8–9	1:7–8	$1:5\frac{1}{2}–6\frac{1}{2}$
V	1:3:10 to 12	1:4	1:10–12	–	–

Note: In special circumstances, e.g. where soluble salts in the background are likely to cause problems, mixes based on sulphate-resisting Portland cement should be employed.
* With fine or poorly graded sands, the lower volume of sand should be used.

Source: Based on table 1 BS5262:1991

unless used on absolutely rigid backgrounds. Wherever possible the weakest mix should be specified unless exceptional resistance is needed to physical damage. In Table 5.6 only the preferred mixes are given, for a wider range of alternatives the British Standard should be checked. Categories for exposure are as follows:

Severe completely exposed conditions for buildings either in the countryside or on the edges of towns, or at some height above surrounding buildings.

Moderate Protection from the elements may be given by other buildings or adequate eaves details. Buildings in towns or suburbs come into this category.

Sheltered Buildings that are well-protected by other buildings close by, and are in areas of low rainfall.

Table 5.6 Recommended mixes for external renderings

Background	First undercoat		Second undercoat*		Final coat†	
	Designation (see Table 5.5)	Thickness (mm)	Designation (see Table 5.5)	Thickness (mm)	Type	Mix proportions by volume‡ or designation (see Table 5.5)
Strong to moderate	II	8–12	II	6–10	Roughcast	$1:\frac{1}{2}:3:1\frac{1}{2}$
					Buttercoat for drydash	II
					Tyrolean	II
Metal lathing	I	3–6	II	10–14	Roughcast	$1:\frac{1}{2}:3:1\frac{1}{2}$
					Buttercoat for drydash	II
					Tyrolean	II
Moderate to weak	III	8–12	III	6–10	Roughcast	1:1:4:2
					Buttercoat for drydash	III
					Tyrolean	II

Note: The nominal overall thickness (excluding texture) is not normally less than 20 mm.
* For severe exposure, the use of two undercoats is preferred.
† For severe exposure, it is preferred that the finish be thrown or rough textured.
‡ Cement:lime:sand:coarse aggregate.

Source: Based on Table 2 BS 5262:1991

Figure 5.17 Prototype housing, Berrylands, Surbiton (built 1936; home of author 1985). The render is applied over a one-brick-thick wall and uses white cement in the mix with dry-dashed white aggregate chippings. It is in excellent condition. These finishes should not be painted over, as this lead to moisture retention and commits the owner to constant maintenance. There are substantial tiled cill details and the render is bell-mouthed over window heads.

When organizing render work it is advisable to appoint contractors with a knowledge of local conditions and successful specifications in a particular area.

Where possible it is preferable to use the lime/cement mixes which have greater flexibility. If there is concern over the durability of these mixes, then a stronger cement mix should be used for bases, dados, mouldings, etc.

The types of finish which can be obtained are described below.

Wood float   This will give a smooth and flat coat. If the surface is raised by a felt-coated float or treated afterwards with a lambswool roller, the sand aggregate is slightly raised to give a softer surface texture which is better at resisting crazing.

Scraped or textured   After initial hardening the surface is scraped hard with thin metal giving a coarser texture than the wood float finish. This is even better at resisting crazing than the textured wood float technique.

Roughcast   Aggregate is thrown on the wall which has already been rolled with a slurry of cement. This gives a very coarse texture, the size of which can be varied according to the aggregate specified.

Drydash   Small aggregates or crushed rock chips are thrown on to the wet render giving a close textured rough surface of natural material. A good hardwearing finish (Fig. 5.17).

Problems with renders

To avoid cracking, the building should be properly designed for the loads applied and prevailing ground conditions. The substrate to which the render is applied should be firm and unlikely to move *or* movement should be catered for by dividing the work up into panels with separating movement joints. As the renders are mostly cement-based there will be overall shrinkage which will transfer tensile stresses to the material and may cause cracking. If adhesion is good, the cracking will be resisted. Successive layers should be allowed to harden thoroughly to minimize any cracking due to shrinkage of earlier undercoats.

Rendering over mixed backgrounds should be tackled in a similar way to plaster but using 300 mm wide strips of extended metal over building paper or polythene sheeting. If possible, visible joints should be made at these junctions and a proprietary filler or sealant used to act as a movement joint.

There may be salts in the backgrounds to be rendered that could give rise to efflorescence. If possible and especially if the substrate is prone to dampness, sulphate resisting mortars, cements, etc. should be used. If repairing or coating old buildings, the use of a sulphate-resistant cement will not prevent salts already present in the substrate from causing damage by 'subflorescence' (Tables 5.7 and 5.8).

Before application the surface must be wetted as otherwise water will be lost from the mix and it will lose workability. Too great a water loss will result in not enough water being left for the chemical reaction with water, and the mix will be weaker.

Table 5.7 Precautions for rendering onto various types of new backgrounds

Background	Characteristics considered	Recommendations
Clay bricks		
All types	Suction and key	Preparatory treatment required for backgrounds with low suction and poor key
	Frost resistance	Check that bricks are correct designation
	Expansion	Continue movement joints through rendering
MN and FN*	Soluble salts	Use sulphate-resisting Portland cement
OL and ON*	Frost and soluble salts	Cover with lathing before rendering
Concrete and calcium silicate, all types†	Suction and key	Preparatory treatment required for backgrounds with low or high suction and for poor key
	Drying shrinkage	Continue movement joints through rendering
	Strength	Use weak mixes on low strength materials
No-fines concrete	Rain penetration	Rendering needed to prevent rain penetration
Stone		
Fair faces	Suction and key	Preparatory treatment required if low suction and poor key
Rubble	Key	Repoint and leave pointing rough or scratched
Metal lathing	Movement	Provide movement joints
	Stability	Fix firmly, use rich undercoat, allow to cure and dry
	Corrosion	Ensure corrosion resistance sufficient for purpose
Expanded plastics boards	Low strength	Use thin reinforced polymer modified rendering or metal lathing as support for cement:sand rendering
	Low suction	
	Moisture movement	
Mixed materials	Differential movement	Use mix recommended for weaker background
	Differential adhesion	Form movement joint or span junction with reinforcement on isolating layer

* Abbreviations: MN Moderate frost resistance, normal soluble salt content clay bricks.
               FN Frost resistant, normal soluble salt content clay bricks.
               OL Not frost resistant, low soluble salt content clay bricks.
               ON Not frost resistant, normal soluble salt content clay bricks.
† For example, *in-situ* concrete, precast concrete units, concrete blocks and bricks, calcium silicate bricks.

Source: Based on Table 4 BS 5262:1991

Table 5.8 Recommendations for rendering onto contaminated or deteriorated backgrounds

Problem	Recommendations
Cracks	
in background	Determine cause of cracks, remove rendering, repair, isolate and reinforce new rendering as appropriate
in rendering	Inspect for hollowness, cut out hollow areas, widen and fill all but very fine cracks and decorate
Hollowness	Repair any defective details, remove hollow rendering, patch the rendering, re-decorate if necessary
Spalled backgrounds	
bricks	Cut out and replace with new bricks or layers of mortar before rendering
reinforced concrete	Cut back to expose any corroded reinforcement, clean and protect, embed in rich mortar
Eroding surfaces	Abrasive blast to sound material and use a stipple coat or cover with metal lathing before rendering
Painted surfaces	Abrasive blast and if necessary bonding treatment, or cover with metal lathing
Salt contamination	Correct any faults, dry brush or abrade off any salts, then use bonding treatment
Oil splashes	Remove all traces of oil or isolate rendering
Organic growth	Kill with toxic wash and remove by brushing

Source: Based on Table 5 BS 5262:1991

Figure 5.18 Rendering de-bonding from a brickwork wall. There are no raked joints in the brickwork and no satisfactory head detail to the render showing that this is an old wall which was subsequently rendered. A sloping fillet detail would be inadequate although a traditional tile creasing detail would be more pleasing and effective. *Bottom*: This shows that when de-bonding occurs due to water penetration and plant growth between the two materials, the load of the render tilts forward inducing a horizontal brittle fracture before falling in a single piece.

Protection

If possible, it is better to avoid painting to allow free movement of water vapour through the material. Once painted, an owner is committed to lifelong building maintenance. If this has to be done, then moisture-permeable paints should be used. It is preferable to use a render specification with a natural aggregate finish that needs no further attention (Fig. 5.19).

5.4 Plasters

Although plaster may be considered to be a ceramic because it is inorganic and nonmetallic, it should really be thought of as a composite material. According to the components in the mix it will be either a particle composite or a fibre composite. Fibres or particles are used as strengthening or crack-stopping elements, depending on the particular range of properties needed in the particular situation. The nature of the composite used has also changed historically, according to the materials available or most economically viable at a particular time.

Figure 5.19 Cinema, Isle of Sheppey, Sheerness. Three dimensional render detailing on a cinema. Although a powerful feature, access is needed for maintenance and a specification for natural aggregates and cement that did not require painting would be more appropriate.

Plaster   The definition of plaster is set out in BS 6110:1984 subsection 6.6.2 (1990). Plaster is a mix based on a binder of lime, cement or gypsum plaster with or without the addition of aggregate, hair or other materials. After the addition of water, it is applied while plastic and it hardens to obtain a surface finish.

***In situ* plastering**   Range of operations by which *in situ* plasterwork is formed.

Suction   Suction is the ability of a *plastering background* or rendering background to absorb water from plaster or rendering respectively.

Care must always be taken to ensure that the water needed for the correct mixing and setting ratio is not unduly absorbed by the background, hence the practice of wetting walls prior to plastering. In order that the wetting is not excessive the use of a bonding agent is sometimes more satisfactory if only to seal some of the voids.

Adhesion   Adhesion is the bond between the two coats or between a coat and a background secured other than by localized mechanical keys.

Bond   The interface strength resulting from adhesion, from a mechanical key or both.

Perlite aggregate   A lightweight aggregate produced from a siliceous volcanic glass when expanded by heat. Particles of perlite can be easily squeezed between the fingers.

Vermiculite   A lightweight aggregate produced from vermiculite (a micaceous material) when exfoliated by heat. (Small particles can be seen to shimmer on close inspection).

Plaster as a composite   Early *particle* composites would include mixes containing a variety of aggregates (sand or ground brick) in a matrix of gypsum or lime. Early *fibre* composites contained straw, hair, or (on a much larger scale) a backing of woven wattle construction or timber lathing on stud frames.

Common particle composites today still include sand but are more likely to include particles from blast furnace waste, or lightweight aggregates such as perlite or vermiculite. Fibre composites may include glass fibres and organic fibres which may be modern polymers or natural cellulose fibres. On a larger scale we have replaced light timber woven or lapping systems with expanded metal and ribbed constructions.

Plasterboard is also a true lamellar composite of gypsum and thick paper.

All of these plaster composites make for a stable plaster finish that can resist some movement and certainly take up a greater degree of applied stress than neat plaster.

Particle composites have crack-stopping capabilities, and fibre composites have directional strength. Early wattle systems and lathe backing on timber studding have a configuration of fibre reinforcement running in two directions, which stabilized an applied matrix. Grid systems with directional reinforcement in two or more directions give added rigidity by equal and opposite support. This is why care should be taken to ensure that modern expanded metal backings have adequate restraint in two directions. This is normally achieved by regular centres of fixings. Ribbed expanded metal is preferable and should be used wherever possible. Early Hyrib expanded metal lathing products recommended Portland cement mixes which still contained a measure of small-scale fibre reinforcement in the form of ox hair, recognizing that additional reinforcement was necessary.

Specification   Although the physical parameters may be clear for making a decision about what type of plaster to use, taking into account internal or external exposure, degree of moisture present, special requirements for sound, fire, hardness, thermal insulation and type of background, there are situations where a decision must be made on an *appropriate* solution for the work. As the use of lightweight plasters becomes more widespread and it is expedient to use plasterboard, it becomes more difficult to consider for each application the use of sand, lime, cement mixes, and also the method of compositional reinforcement. In restoration work, wherever possible, plaster specifications should match the original behaviour will then be similar and stress-cracking between old and new controlled to some extent.

The above principles apply to internal and external situations. The chief difference between internal and external finishes relates to exposure and determines the matrix to be used. All calcium sulphate minerals (gypsum) tend to dissolve very slightly in water and, in effect, undergo hydration reverting to the original salt of the gypsum mineral with a consequent increase in volume. They are are suitable for use externally where it is more appropriate to use cement:sand and cement:lime:sand mixes (see section 5.3 on Renders). Cement goes through an irreversible chemical reaction with water and is generally regarded as insoluble and more stable.

Gypsum: mineral origin   Plaster comes from the mineral gypsum a naturally occurring hydrated calcium sulphate (chemical formula of $CaSO_4.2H_2O$). It comes in several different colours: white, grey, red or yellowish brown. There is also a colourless version. It has a low Mohs hardness of 2.0 and can be scratched with a fingernail.

For advice on the application of plasters it is necessary to use the Codes of Practice. In the case of internal plastering use BS 5492:1990 and for external plastering use BS 5262:1991. Both codes deal with the materials available, the properties of the backgrounds, and how specific plasters should be selected and used. See also BS 8000 Part 10:1989 *Workmanship on Building Sites: Code of Practice for Plastering and Rendering*.

Although the British Standard BS 1191 Parts 1 (Gypsum based) and 2 (Pre-mixed, lightweight) sets out four classes of plaster (A, B, C and D), these options are difficult to achieve in practice. The different classes are based on the extent to which the gypsum is heated and the amount of water that is released. It must not be forgotten that when water is driven off in heating, it involves the breaking of chemical bonds and the release of constituents that make up a stable compound. When water is re-introduced at a later stage in mixing, it immediately re-combines with the dehydrated plaster in an exothermic reaction which gives off heat.

If not very much water is driven off in the manufacture of a plaster, then the addition of water will give a fast re-set to the material. Plaster of Paris is an example of a class A plaster which has only been part dehydrated in the manufacturing process. To be workable for site conditions this plaster must have a retarder added to delay the hardening process. When a class A plaster is modified in this way it becomes a class B plaster.

The greater the amount of water driven off in the calcining of gypsum, the slower the reaction when remixed with water. Class C plasters have all the water of crystallization driven off. When remixed with water the reaction may be so slow that it is impractical to use the material unless an accelerator is added to speed up the reaction.

Classes of plaster

Class A Plaster of Paris   Plaster of Paris is a hemi-hydrate with no retarder. It is commonly used for casting running moulds or small areas of filling, etc. In running moulds a pattern is made up in zinc with a stiffening profile board set behind, and the plaster detail is built up gradually with plaster and scrim, with the final details run in neat plaster.

Class B Retarded hemi-hydrate gypsum plaster   A retarded hemi-hydrate contains a retarder that controls the set. These plaster types are used as matrices for lightweight aggregates.

Browning	a1	Thistle browning, thistle slow set, thistle fibred
Metal lathe undercoat	a2	Thistle metal lathing
Final coat	b1	Thistle finish, sirapite B, British gypsum
	b2	Thistle board finish

Class C Anhydrous gypsum plaster   Completely de-hydrated gypsum, incorporating an accelerator to achieve a set in a reasonable time. Because of its hardness this plaster is often used as a final setting coat, originally referred to as 'sirapite'. It is no longer easily available.

Class D Keene's plaster   This completely de-hydrated plaster has a very slow set, but a high workability to achieve a good finish. It was previously used for arrises, dados, skirtings, squash courts and surfaces where a very hard surface was required but is no longer made in quantity. Substitutes containing cement are often reinforced with glass fibre and more readily available.

Standard Keene's

Polar white cement	(fine)
Polar white cement	(standard)

The British Standard 1191 is split into two parts. Part 1 deals with pure gypsum-based products; Part 2 deals with pre-mixed lightweight plasters which contain lightweight aggregates such as vermiculite or expanded perlite. These lightweight plasters are the most commonly used plasters today and are grouped into undercoat and final coat plasters.

Type (a) Undercoat plasters

Browning	Carlite browning
Metal lathing	Carlite metal lathing
Bonding	Carlite bonding
	Carlite welterweight bonding coat plaster

Type (b) Final coat plaster

Finish plaster	Carlite finish
	Limelite finishing plaster

Cement plasters that are now used instead of class D plasters:

Undercoat premixed: Limelite
Resinous for squash courts: Proderite formula S-
                 based screed and finish

When using browning and metal lathing plasters one has some control over the soluble magnesium and salt contents, but with the bonding plaster one does not. This is worth noting where there is a possibility of plasters being used in damp conditions which could encourage efflorescence.

One coat plasters are becoming more popular especially for white plaster. There is a trend towards the production of a greater range of plasters that introduce many more components to the traditional gypsum matrix. Although these can cope with a great variety of situations it means that the basis for specification is becoming more product-based. Specifiers will not be able to quote the British Standards but should satisfy themselves that a chosen product will comply to their requirements in terms of background compatibility, durability, fire-resistance and acoustic control.

Using the British Standard BS 1191 The standard gives a good checklist for planning plastering and also for making sure that points of detailed design are considered, e.g. with regard to walls and frames, corners, covers, openings and architraves. The sequence of trades is critical: all services should have had their first fix and all openings should be properly framed up to provide an edge for finished work. Depending on the plaster mix used, care must be taken to ensure the programming allows for full drying and hardening prior to final decoration.

There are additional types of plaster for use in specific situations mentioned in the standard:

Thin wall plasters (skim and fillers). These often contain organic binders and may not be compatible with the substrate particularly if *damp*. They should always be checked for compatibility and the manufacturers instructions carefully followed.
Projection plasters. These are sprayed finishes.
X-ray protection plasters. This plaster contains barium sulphate as an aggregate. Barium is a heavy metal whose salts are used for protection against radiation.
Acoustic plasters. These contain porous, sound-absorbing aggregates. Care should be taken that the acoustic properties are not ruined by indiscriminate painting that may clog up the fissures and

holes which affect sound distribution.
Cement plasters. These contain cement and lime, and lightweight aggregate.

When cements are referred to in the standard, it should be noted that Portland cement is to BS 12:1991, Portland blast-furnace cement to BS 146:1991 (now revised in terms of EN196 test procedures), and masonry cement to BS 5224:1976. There are references to high-alumina cement which must be used strictly in accordance with the building regulations and with users well aware of the problems in wider applications. (See BRE Information Paper 22, 1981 *Assessment of chemical attack of high-alumina cement concrete*.) Normal aggregates used, e.g. sand, should be to BS 1198:1976, BS 1199 and BS 1200:1976.

The reason for specifying all components to British Standards is that they do ensure a basic quality and hence purity of the material. There are many unknowns in terms of the exact chemical composition or contamination in the substrates, and known components give an assurance of chemical and physical stability. Problems can arise involving efflorescence, staining, unpredictable setting, etc. *Even water should be specified to BS 3149:1980.* Bonding agents are to BS 5270:1989. As they coat porous material and tend to diminish the size of pores at the same time, they will also reduce the effects of suction, the take up of moisture by capillary action. Scrim (a loose weave hessian strip) is used to reinforce corners and joints as needed. It is normal practice to plaster up corners first and then plaster walls up to the junction to minimize the problem of cracking.

Characteristics of different mixes of plaster
Plasters contain a variety of components. The mix chosen will have specific characteristics.

Cement/lime/sand These mixes are highly workable due to the lime component, but they develop their strength slowly, often forming a hard outer skin which increases in strength over the years. Hardening is by carbonation which is why it is so difficult to match exactly old plaster specifications. Each coat has to be allowed to harden first and should be laid on with a wooden float: metal trowelling will produce fine cracks as surface crazing occurs on a cement slurry. The wood float slightly raises the texture and the sand particles act as crack stoppers to surface cracking.

Lime and gypsum plasters These are highly workable mixes and they are relatively stable. As the lime component shrinks it is compensated slightly by the expansion of gypsum. The presence of lime gives an alkaline environment which inhibits the corrosion of metals.

Lightweight cements   These are similar to ordinary cement in use and they can be finished with a lightweight finishing coat of cement or a coat of gypsum.

Gypsum plasters   These are all workable plasters and are mostly class B. They all expand on setting, and this makes some kinds of work more difficult, e.g. running cornices (which normally use class A plaster): if the work is not carried out quickly, the running mould will snag and cause unevenness. If very smooth surface finishes are wanted then class C and D plasters should be used. If worked on, a finish similar to polished marble can be achieved.

Some gypsum plasters contain accelerators or frost-proofing additives. These compounds tend to be salts (see glossary in section 4.2) with active ions which can initiate corrosion in metals.

General properties   Gypsum plasters generally have an insignificant effect on thermal transmittance, condensation, sound absorption (except acoustic plasters) or sound insulation (which is a function of mass). They are extremely good fire-retardants and are classified as class O (non-combustible). As a material, gypsum plaster cannot degrade until all the water of crystallization is driven off, and the temperature of the material remains at about 100 °C until this happens. This delay in the rise in temperature may give enough time for escape from a fire.

Substrates

BS 1191 identifies the following problems in plastering:

(1) Substrates: inadequate key weakness or contamination with dirt and grease or water   Any of the above faults in the substrate will result in failure in bonding between plaster and substrate, and the plaster will move, fall off or flake.

(2) Premature drying   If the plaster has started to harden before application to the background the plaster will have insufficient moisture left for adhesion, and will fall away.

(3) Salt formation and interference   It is likely that

salts will be present in substrates, particularly in brickwork and in mortar joints. If the substrate is wet, the salts will continually leach out until the source of moisture disappears or the salt content is exhausted. It is important to check substrates for efflorescence and program work accordingly.

(4) Structural movement, moisture movement and thermal movement   Movement will cause cracking and could lead eventually to total bond failure, and the plaster will shear away from the substrate. If a substrate is still hardening, especially if cementitious materials are being used, it will be shrinking, and this may cause failure.

(5) Inadequate suction control   If a great deal of water is lost from a mix during plastering then the outer surface will be embrittled and more likely to fracture with hairline crazing. Sometimes differential moisture loss can be seen by the patterning of surfaces behind the plaster which are said to 'grin' through. Hydration discoloration may be seen quite clearly where a different amount of moisture is lost from the mix into different adjacent backgrounds, for example, brickwork and mortar joints. The amount of water used in the hydration reaction will affect the final tone of the set plaster.

(6) Any combination of (1) to (5)   If a failure occurs it is quite common to look for one factor which is easily identifiable, and that will of course lead to one clear remedy. In fact gypsum can take a good deal of stress and it is more common to find that there are several factors involved in failure.

Specification   In choosing a specification for plastering the BRE Information paper No 213 (May 1978) should be referred to. The exact specification will depend on the background to be plastered and the final finish needed.

Types of backgrounds

One can plaster a variety of backgrounds from concrete to polystyrene, and from rough to smooth. The successful bonding of plaster on to a background relies on some degree of absorption, and not just on mechanical keying (see section 3.3 on bonding). Bonding cannot be estimated visually. If there is any doubt as to the mix specified and to the background to be treated, then a sample panel should be built (especially if it involves a departure from the combined experience of contractor and specifier). If there is trouble with adhesion then bonding agents may well be needed. This is particularly true with concrete, where the density of mix and aggregates is such that there is no absorbency.

Solid backgrounds
Dense, strong, smooth backgrounds include:

High density concrete, brickwork and blockwork In order for plaster to set, the substrate must have some suction. As density correlates with strength it should be expected that the strongest materials with the highest compressive strengths will be difficult to plaster successfully.

Moderately strong and porous materials include:

Most bricks and blocks The increase in porosity will give reasonable adhesion and mechanical keys.

Weak and porous materials include:

Lightweight concrete and blocks The adhesion is increasingly good although plaster may lose so much water into the body of the material that bonding agents may be applied to the substrate to reduce pore sizes.

'No fines' concrete includes:

Concrete which shows gap grading where the fine aggregate is partly omitted The porosity of this material is so high that mechanical adhesion is a greater factor than in all the other materials in keeping the plaster on the surface.

Sometimes the plaster has to bridge over different materials. There could be junctions between concrete and brick, brick and blockwork, or over studding. These will need a reinforcing bridge, usually of expanded metal. The expanded metal should be fixed over building paper before plastering. The lack of adhesion to the real substrate behind the expanded metal will minimize cracking.

It used to be common practice to make straight cuts on some junctions, particularly in corners: this makes better sense than attempting to bridge over dissimilar materials. When walls were more commonly papered it didn't matter so much and was an obvious solution for butting a lathe and studding wall at right angles against brickwork. For junctions where a studding wall is running with a brick wall it was an easy matter to carry over the timber lathes onto the brickwork. One useful nineteenth-century detail often forgotten now is to use a rounded moulded detail in timber on external corner details and plaster can butt up to the moulding with a bevelled finish.

Slab backgrounds These include woodwool and compressed straw boards.

Boards These include plasterboard and expanded polystyrene.

The Greater London Council Bulletin No 100 gave specific advice on plastering over polystyrene. It is a material that appears to be unsuitable as a background, because it is smooth with no porosity and presents no surface topography that would give an adequate mechanical key. The only way of plastering successfully is to use a plaster designed to supply its own means of adhesion.

This bulletin discusses Cartlite metal lathing plasters, Carlite welterweight bonding coat and Thistle projection plaster. It is not advised to used PVA bonding agents. After all, the plasters have been designed on the assumption of no bonding capability of the background and to ignore that fact will lead to a poorer specification, not a better one. Table 5.9 gives recommended thicknesses for plastering on polystyrene.

Metal lathing and decking

Painted and tiled backgrounds Although the standard advises on the use of a bonding agent before plastering, it is not a practice to be advocated. The finish will rely on the structural strength of the interface between paint and wall or tile and wall.

Table 5.9 Recommended thicknesses for plastering onto polystyrene

Plaster	Thickness of plaster (mm)			
	Ceilings		Walls	
	Undercoat	Finish	Undercoat	Finish
Carlite bonding	8	2	11	2
Carlite welterweight	8	2	11	2
Thistle	10		13	

There may also be problems with compatibility between bonding agents and paint films.

Bonding agents

The use of bonding agents is well known. They consist of organic polymer-based materials containing styrene butadiene, polyvinyl acetate or acrylic resins. Polyvinyl acetate is often shorted to *PVA*. Using the recommendations by 'Febond' for their market bonding agent, it should be used as a two-part process. The wall to be prepared for plastering should first be sealed to reduce porosity, and then the coat that assists bonding is applied. The use of PVA in wet conditions is not encouraged and the material should also be protected from frost. Table 5.10 gives instructions for the use of bonding agents. (Note: Unibond have an Agrément Certificate for their PVA bonding solution, No 1741:1986.)

There is also an Agrément Certificate for *EVA* (unibond 1741:1986). EVA is described as a modified one-part ethylene-vinyl acetate emulsion. It is for use with gypsum plasters 'in dry service conditions only', but can be used with sand and cement mixes in wet service conditions. It does 'significantly' alter water vapour transmissions. It contains 55% solids and can be mixed with water or with ordinary Portland cement and water. Durability is given as 'not less than 10 years'. As these products are relatively new the long-term performance of these bonding agents has not yet been tested. It should be remembered that the service operating temperatures for many polymers is critical, and, in extreme cold, the polymer may transform into a glassy state and become embrittled.

Instructions for using bonding agents should be followed carefully. They should not be used neat and the background is normally primed with a diluted solution.

Ethylene-vinyl acetate (EVA)

Dry conditions   A first priming coat should be diluted 1:5 with water and allowed to dry. The second coat should be 1:3.

Wet conditions   A priming coat of 1:5 should be used, as for dry conditions, and allowed to dry, followed by a slurry of 1 part EVA to 2 parts OPC (ordinary Portland cement) applied while the coat is still wet.

Plaster thicknesses

The Gypsum Products Development Association recommend the use of PVAC bonding agents for par-

Table 5.10 Use of bonding agents

Application	PVA:water ratio	
	Porous surfaces	Semi- and non-porous
Seal	1:4	1:6
Bond	1:2	1:2

ticular situations, including the application of a final coat of gypsum plaster (2 mm thickness) to concrete, and for heavier coatings of plaster to plasterboard. Specific applications can always be checked with the manufacturers first.

Walls   One- and two-coat work is now common practice for cement and gypsum depending on the type of coating used. The final thickness will be 13 mm for brick, block or metal lathing. On concrete the maximum thickness should be 10 mm unless the surface is treated to give a deep mechanical key. As the adhesion on concrete is relatively low compared to brickwork, the load of plaster that can be supported is less. Three-coat work will give thicknesses approaching 19 mm, and used to be even greater when the type of background warranted a build up of coats to take out great differences in level. Thick coats were more common where the degree of mechanial interlocking was deeper, e.g. on timber lathes.

Ceilings   On soffits there are more acute problems connected with defying gravity. Old methods for plastering ceilings worked well with initial coats drawn at 45 degrees over lathes so the plaster would curl over like miniature waves and mechanically key over the timber. Thicknesses could then build up to 25 mm. Today with the lack of a mechanical adhesion, plastering on soffits cannot go over 10 mm in thickness and is usually plasterboard.

Plasterboard   The basic thickness of plaster on gypsum plasterboard should not be less than 5 mm although there is a tendency for 'skim' coats to mean just that, i.e. to be inadequate. Plaster can be applied up to 10 mm thick on plasterboard as the adhesion with the card face is excellent and two-coat work is recommended. The plasters used should comply with BS 1191 Part 1, Class B for single coat work and type A3 for undercoats in two-coat work. Note that the surface must not be wetted prior to plastering.

Application

Backgrounds with large voids or uneven surfaces should be dubbed out (levelled) before a first coat of plaster is applied. The first coat or undercoat should be applied over a background that may need preparation by wetting, or treatment with a bonding agent. After surface preparation the undercoat should be *scratch-keyed* with a downward action in preparation for the next coat and allowed to harden. Second coats should also be scratch-keyed for the final coat if a three-coat system is used.

Decorative plasters

Textured finishes There are one-coat treatments that have a possibility of many different textures; for example *Artex, Wondertex* and *Pyrok* (which are designed for different conditions) and internal and external finishes with high insulation qualities.

White plasters These are often used as one-coat plasters, known as projection plasters, machine applied with a resin build and ready for decoration; for example *Snowplast* and *KIPS Goldband* or *Whiteband* (finishing plaster).

Universal plasters These are used as one-coat finishes and are often a calcium sulphate hemi-hydrate plaster with lightweight aggregate; for example *Thistle Universal One-coat* or *Lafrage Ivory Finish Plaster*.

Hardwall plasters These are designed to have high impact resistance. They are usually a retarded hemi-hydrate plaster with perlite and vermiculite aggregates; for example *Thistle Hardwall or KIPS Blackband*.

Renovating plasters These plasters are designed for older buildings which often have damp conditions. Additives promote early surface hardening and development of strength despite moist conditions; fungicides are often included. They should not be used below ground level. They include *Thistle Renovating* and *KIPS Renovating*.

Care of plaster on site

Gypsum plaster and cements are both vulnerable to the absorption of moisture. In their bagged and powdered state, they will readily re-combine with water. If damp plaster is mixed up, it goes off much faster than usual in a so-called 'flash set', and may be impossible to use. This explains the need to keep all these materials tightly bagged, under cover and off the ground. They must be used in strict order of delivery to site. For the same reason, when working on site, fresh plaster should not be added to earlier mixed batches that have been left standing. Plasterers' tools should be cleaned with every new batch made up, otherwise the plaster mix will go off as hardening is accelerated.

Cement and gypsum

These should not be used together in the same wet mix. Cement and gypsum react chemically, causing great expansion as sulphate attacks calcium oxide. Some proprietary skim treatments are cementitious and can also give rise to chemical reactions and sometimes not set properly if applied to damp gypsum bonding coats. More rarely, trace iron sulphides in cement mixes may give off hydrogen sulphide (pungent smelling) if used in combination with slightly acid preparations. Compatibility between plastering components must be checked and adhered to.

Stucco

The definition of stucco used to be very general, but common stucco for external use consisted of three parts clean sharp sand to one part of hydraulic lime. If a coarser appearance was needed (to imitate stone) then a larger grained sand would be used. This mix would be spread thinly over the backing coat (which was not fully set) with a felt-covered hand float which lifted the particles up to give the final texture.

Ornamental plasters

These plasters were intended to imitate marbles. The following descriptions of scagiola and marezzo marble come from Rivington's *Notes on Building Construction* Longman, Green & Co. 1901, Vol III, p.250.

   'Scagiola is a coating applied to walls, columns, etc. to imitate marble. It is made of plaster of Paris, mixed with various colouring matters dissolved in glue or ising glass; also with fragments of alabaster or coloured cement interspersed throughout the body of the plaster.'

   (Note: Ising glass is a chemical term for fishglue, and is prepared from fish bladders. It was originally used as an adhesive, but is now more widely used in the food and drink industries.)

   This mixture would be applied over an initial coat of lime and hair mortar which is described as having been 'pricked up' and allowed to harden fully. This first coat would be set over a traditional lathe construction on timber framing that might make up a pilaster detail. The scagiola surface application would be finished by rubbing with a wet pumice stone when dry, rubbing

over with 'tripoli' and charcoal, and polishing with a felt rubber of tripoli and oil. It would be finally wiped with oil. 'Tripoli' probably refers to the mineral tripolite which is a variety of opaline silica, and must have been used as a fine finished abrasive. *Marezzo marble* is also a kind of plaster made to imitate marble.

'A sheet of plate glass is first procured, upon which are placed threads of floss silk, which have been dipped into the veining colours previously mixed to a semi-fluid state with plaster of Paris. Upon the experience and skill of the workman in placing this coloured silk the success of the material produced depends. When the various tints and shades required have been put on the glass, the body colour of the marble to be imitated is put on by hand. At this stage the silk is withdrawn, and leaves behind sufficient of the colouring matter with which it was saturated to form the veinings and the makings of the marble. Dry plaster of Paris is now sprinkled over to take up the excess of moisture, and to give the plaster the proper consistence. A canvas backing is applied to strengthen the thin coat of plaster, which is followed by cement to any desired thickness; the slab is then removed from the glass and polished.'

Alternative finish:skim coat and embodied pigments   Instead of relying on paint finishes to plaster walls, if a permanent finish is wanted with body colour, a skim coat of plaster can have integral colour as a complete finish. It is better to use white plasters such as snowplast, and add either artists' pigments or acrylic colour. Even mixing will give homogeneous colours, and the addition of PVA bonding agent will prevent dusting and help with adhesion on old backgrounds. It is an ideal solution for refurbishment if colours can be chosen for a long life as it obviates the need for extensive filling and patching if one coat proprietry mixes are used. Sample areas should be done first to gauge proportions. As a guide, a 5 litre bucket with $\frac{1}{2}$ cup of PVA and either 5 teaspoons of pigment or 5 inch squirts of acrylic colour will give a reasonable finish and body colour.

Plasterboards

Plasterboard is used on timber and metal studding construction systems. It is also a fast-finishing wall technique for masonry, giving an air space which improves thermal insulation.

When used as a vertical or horizontal lining, decoration can be applied directly to the plasterboard after fixing and treatment of joints or one board can be used as a base for gypsum plasters. Plasterboard is now used extensively to help upgrade the fire resistance and sound performance of new and existing floor and framing systems. There is now a large supply of the mineral gypsum (used in making plasterboards) since it is a by-product of the de-sulphurization processes now used in power stations.

The British Standard for plasterboard (BS 1230 Part 1:1985) includes test procedures. Plasterboard consists of a core bonded to paper liners which can be different grades depending on whether the surface is to be plastered or decorated. Most boards are made with a grey side for plastering and an ivory side for decoration. Boards are usually delivered in pairs with the protected internal sides ready for decoration.

Plasterboards are also made with backings of foil and other laminates such as a polystyrene (extruded or expanded), polyurethane and phenolic foam as a vapour check and mineral wool as insulation.

Plasterboard is a true composite. The strength of plasterboard relies on the card surfaces being intact to resist stress, so damage incurred in fixing will weaken the material. Nails should pierce but the heads should just touch the surface and not puncture the paper.

Baseboard is used as a base for plastering. The long edges are rounded or bevelled giving a key without the need for applying scrim. For a good finish with scrimmed joints, tapered-edge boards should be used. Wallboard and baseboard can have improved fire protection performance and are denoted by the suffix F.

A variety of properties and applications of plasterboards, baseboard, and wallboards are given in Tables 5.11–15. For greater detail about properties and sizes of individual plasterboard types, or specifications to suit particular applications, contact the Gypsum Products Development Association.

Joints

Proprietary joint fillers and tapes for the pre-formed edges of plasterboard are sponged over after being filled, ready for decoration. If edges are to be cut they should be chamfered first to allow for filling; using a knife is more likely to take chunks out of the edge than using a surform blade.

Fixings

Nails for fixing plasterboard should be galvanized and preferably have a profiled shank to prevent 'pull-out'. Centres for fixing should be every 150 mm for ceilings and every 200 mm for walls. Nails should be positioned at least 13 mm into the board otherwise the plasterboard will shear at the edge. Manufacturers

Table 5.11 Basic dimensions of plasterboard

Dimension	Available sizes (mm)	Dimensional tolerances
Width	600, 900, 1200	±5 mm
Length	1800, 1829, 2286, 2350, 2438, 3000, 3300, 3600	±6 mm
Thickness	9.5, 12.5, 15, 19, 25	±0.5 mm for thicknesses ≤9.5 mm; ±0.6 mm for thicknesses >9.5 mm

Note: Plasterboard has to conform to specified breaking loads; its strength is not equal in both directions since the paper used has non-uniform directional strength owing to its manufacture.

Source: Based on Section 5.11 BS 1230 Part 1:1985

Table 5.12 Wallboard breaking loads

Board thickness (mm)	Minimum breaking load (N)	
	Transverse	Longitudinal
9.5	170	405
12.5	230	535
15	260	620
19	305	765
25	380	1000

Source: Based on Table 2 BS 1230 Part 1:1985

Table 5.13 Baseboard breaking loads

Board thickness (mm)	Minimum breaking load (N)	
	Transverse	Longitudinal
9.5	125	180
12.5	165	135

Note: Water absorption should not be less than 5% and the baseboard should comply with class 1 surface spread of flame.

Source: Based on Table 3 BS 1230 Part 1:1985

Table 5.14 Wallboard and baseboard applications

Type	Description	Application
1	Gypsum wallboard	Walls/ceilings (decoration)
2	Gypsum base wallboard	For veneers
3	Gypsum moisture-resistant wallboard	Moisture risk (general)
4	Gypsum moisture-resistant wallboard	Moisture risk (surface)
5	Gypsum fire-resistant wallboard	As type 1 but improved fire resistance
6	Gypsum baseboard	For plastering
7	Gypsum fire-resistant baseboard	As type 6 but improved fire resistance

Note: Refer to subsection 6.6.2 of BS 6100:1990 for a more extensive list of plasterboard types.

Source: Based on Table 1 BS 1230 Part 1:1985

recommendations should always be followed for fixings for plasterboards. Tools for fitting and nailing are specialized to minimize damage to the board edges and surface. (See the Gypsum Handbook for detailed advice).

Renders

Renders are now associated with failure in many instances despite a long history of a successful use in traditional building. This is due to a complete change in the materials used to achieve a homogeneous but relatively thin coating over the years. There is also a change in expectations with regard to the standard of finish required, and the optimization of physical

Table 5.15 Framing for boards

Framing	Board thickness (mm)	Centres (mm)
Vertical		
Lath	9.5	450
Baseboard	9.5	450
Lath	12.7	600
Plank	19	800
Horizontal		
Lath	9.5	400
Baseboard	9.5	400
Lath	12.7	450
Plank	19	750

Source: Based on Tables 7, 8 and 9 BS 8212 Part 1:1988

dimensions. The surface area to be covered is so large that problems of movement, whether by stress or by thermal response and flexure, must be catered for in a way which will show on a building facade.

Render is also expected to be successful on a variety of substrates, from concrete to brick, to block, and even over framing systems. Render is often erroneously thought of as a material in its own right which is simply applied over all these different types of construction. Every time the substrate changes, the nature of the render coating used must also change. Render and substrate must be thought of as a composite material and compatible in terms of likely physical behaviour under stress – after all, both skins must behave as one.

Sources of materials   In England the most extensive resources for gypsum are in the North of England. Hydrous (gypsum) and anhydrous (anhydrite) forms of calcium sulphate are found in West Cumberland, the Vale of Eden and South West Durham. Gypsum is usually found close to the surface with anydydrite lower down. Although anhydrite is mainly used to make ammonium sulphate fertilizer and sulphuric acid, there is a cement by-product. In the *British Regional Geology* series (HMSO, 1971) it was estimated that two million tons of anhydrite and gypsum was produced annually from this region. There are other minor deposits in East Leake, Leicestershire, and Chellaston, Derbyshire, also Fauld near Tetbury in Staffordshire. The ornamental alabaster of Chellaston and Fauld is a translucent form of gypsum. Tables 5.16–20 give a variety of properties of various plaster mixes and substrates.

Refurbishment and replacement

Many buildings undergoing refurbishment require new plaster, especially if the building fabric has been subjected to damp. There is no point replastering unless any structural movement has been stabilized or sources of damp dealt with, and it is better to cut out cracks and remove all loose plaster for better bonding. To optimize performance, it might be preferable to use lightweight plasters to improve thermal performance, or heavyweight plasters to improve sound insulation. It could be relatively easy to specify lightweight plasters on external walls and heavyweight plasters on flank or party walls. Where possible, replacement plasters should also be in character with the existing buildings, and the normal use of beads should be considered where round-edged details using timber mouldings is appropriate. The BRE Good Building Guide No 7 Jan 1991 *Replacing Failed Plasterwork* gives useful advice and checklists.

5.5 Stone

For the use of stone generally, its composition, characteristics and sources, see MBS: *Materials*, Chapter 4.

Stone is not regarded today as a major structural building material. Most stone is won from quarries to supply aggregate material for concrete or hardcore for the refining of steel, and for the chemical industry. Many quarries have fallen into disuse, and when masonry is designed and detailed in walling and columnar structures, it is often to repair existing buildings. Consequently, in new work, stone is now far more commonly used as a cladding material. The technology of using stone in this way depends entirely on adequate methods of fixing, and careful detailing to ensure that weather-resistant joints are achieved. The detailing and dimensioning on working drawings should include realistic dimensional tolerances.

Stone will vary according to its geological source. Any mineral suddenly exposed to the atmosphere may be unstable. Most minerals were formed in the Earth's

Table 5.16 Cement-based plaster mixes

Mix designation	Cement:lime: sand	Cement:ready-mixed lime:sand		Cement:sand (using an air entraining admixture)	Masonry cement:sand
		Ready-mixed lime:sand	Cement: ready-mixed material		
i	$1:\frac{1}{4}:3$	1:12	1:3	–	–
ii	$1:\frac{1}{2}:4-4\frac{1}{2}$	1:8–9	$1:4-4\frac{1}{2}$	1:3–4	$1:2\frac{1}{2}-3\frac{1}{2}$
iii	1:1:5–6	1:6	1:5–6	1:5–6	1:4–5
iv	1:2:8–9	$1:4\frac{1}{2}$	1:8–9	1:7–8	$1:5\frac{1}{2}-6\frac{1}{2}$

Source: Based on Table 3 BS 5492:1990

Table 5.17 Characteristics of plaster undercoats

Plasters	Shrinkage or expansion	Strength and hardness	Remarks
Lime:sand 1:2–3		Weak and soft	Takes a long time to harden and is not recommended for present day practice
Cement-based undercoats Cement:sand 1:3–4	All shrink on drying. Too much clay or fine material or sands of uniform particle size make for high shrinkage. Contents of clay or fine material should not exceed 5%. Strong mixes tend to develop a few large cracks; weaker mixes develop finer and distributed cracks	Strong and hard	Hardens progressively. Undercoats should be allowed to dry and shrink before applying the next coat. A workability aid or the addition of one-quarter to one-third volume of lime helps application
Aerated cement:sand 1:5–6 1:7–8 Cement:lime:sand 1:1:5–6 1:2:8–9 Masonry cement:sand 1:4–5 1:5$\frac{1}{2}$–6$\frac{1}{2}$		Sufficiently strong and hard for most purposes, although neither the 1:8 aerated cement:sand, the 1:2:9 cement:lime:sand nor the 1:6 masonry cement:sand mixes are sufficiently strong to receive neat gypsum plaster finishes	Hardens progressively. Undercoats should be allowed to dry and shrink before applying the next coat. The aerated mixes do not develop high suction on drying, and they are more effective barriers to the passage of moisture and efflorescent salts
Premixed lightweight cement undercoat including renovating	Shrinks on drying	Sufficiently strong and hard for most purposes but the manufacturers should be consulted as to its suitability for use with particular gypsum final coats	Hardens progressively. Undercoats should be allowed to dry and shrink before applying the next coat. They are more effective barriers to the passage of moisture and efflorescent salts
Premixed lightweight gypsum undercoats		Varies with the type of undercoat. Although they are more liable to indentation than cement-based mixes, their resilience tends to prevent more serious damage	Sets quickly. It is essential to use the undercoat appropriate to the background. On high suction backgrounds high suction browning grade will assist application
Projection plasters and One-coat plasters	Their relatively small movement is easily restrained by the backings	Sufficiently strong and hard for most purposes	Offers the advantages of an equivalent two-coat plaster system on suitable substrates
Premixed gypsum plasters and gypsum renovation plaster			Sets quickly. Strength continues to develop over several days

Source: Based on Table 1 BS 5492:1990

Table 5.18 Characteristics of final coats

Plasters	Shrinkage or expansion	Strength and hardness	Remarks
Lime plasters Neat or sanded (mix proportions by volume 1:0–1)	Shrinks on drying. The addition of well graded fine clean sand progressively reduces this shrinkage	Weak and soft, easily damaged	Takes a long time to harden and is initially suitable only for a permeable decorative finish that is not affected by alkali
Gypsum plasters Class B*	Expands as it sets although this expansion is extremely small. Subsequent movements are small. Too rapid drying may lead to delayed expansion	Hard	Requires a strong undercoat. Predetermined setting time
Thin-coat plasters	Similar to class B	Slightly lower than class B	Requires a strong undercoat. Gradual setting characteristics. Quick drying. Suitable to receive most types of decoration
Premixed lightweight plasters (final coat)	Similar to class B	Surface hardness similar to neat class B gypsum plaster. Liability to indentation varies with the type of lightweight undercoat used, but the resilience of these tends to prevent serious damage	Requires a strong undercoat. Setting and drying characteristics as for neat class B gypsum plaster
Projection plasters and One-coat plasters	Similar to class B	Hard	Requires a strong undercoat. Gradual setting characteristics. Normal drying. Suitable to receive most types of decoration
Renovation finish plasters	Similar to class B	Hard	Requires a strong undercoat. Setting and drying characteristics as for neat class B gypsum plaster

* A small proportion of lime, not exceeding $\frac{1}{4}$ part by volume, may be added to certain class B plasters.

Source: Based on Table 2 BS 5492:1990

crust in the absence of oxygen. Any new face presented to the air after quarrying will change, and so it needs 'seasoning' before use. Some stones when quarried are released from constant pressure, so the crystal lattices of the minerals sometimes expand. Feldspars and ferromagnesian silicates absorb water and over a long period of time (hundreds of years) can decay into clays. Carbonate and sulphate minerals dissolve very slightly in water. This is a continual although slow process. In rain, (which is slightly acid) calcium carbonate will convert to the more soluble calcium bicarbonate, hence the erosion of limestone buildings. This erosion has been accelerated by the more corrosive nature of the atmosphere in an industrial society, sulphur dioxide and carbon dioxide dissolved in water inflict damage faster as they are acid solutions. Even salt from the sea and chlorides in industrial atmospheres can be converted to weak solutions of hydrochloric acid and dissolve carbonate rocks. Nitric acid is produced from oxides of nitrogen, natural products from the combustion of fossil fuels. Carbonic acid is produced when carbon dioxide dissolves in rainwater. All of these products form powerful corrosive solutions contributing to decay. Even salt used on roads and pavements in winters is a large source of chlorides that can attack the stone bases

Table 5.19 Backgrounds: characteristics and suitable plastering systems

Background		Suction	Key or bond	Drying shrinkage	Type of finish*	Suitable alternative undercoat plasters			Remarks
						Premixed lightweight plasters		Undercoat based on cement (for mix designation see Table 5.16)	
Class	Type					Gypsum	Cement		
Solid	Normal clay brickwork and clay blockwork	Moderate to high	Good if joints well raked or bricks keyed	Negligible	I	–		ii, iii or iv	Background should be dry to minimize efflorescence
					II	Browning	Backing or browning	ii, iii or iv	
					III	None required			
					IV	None required			
	Dense clay brickwork (other than engineering brickwork), calcium silicate brickwork, concrete blockwork and concrete brickwork	Low to moderate	Variable. Bonding treatment may be necessary for cement-based undercoats	Low to high for calcium silicate brickwork. Low to moderate for concrete blockwork and concrete brickwork. Negligible for dense clay brickwork	I	–		ii, iii or iv	Background should be dry to minimize shrinkage except for dense clay brickwork
					II	Browning or bonding	Backing or browning	ii, iii or iv	
					III	None required			
					IV	None required			
	Clay engineering brickwork, dense concrete	Low	Baking joints, hacking, spatterdash or bonding treatment may be necessary. With cement-based plasters bonding is essential	Negligible for clay engineering brickwork. Low to high for dense concrete	I	–		ii, iii or iv	Sufficiently flat concrete surfaces may be plastered with thin-coat plasters or board finish plasters; bonding treatment may be necessary
					II	Bonding	Backing or browning with bonding treatment	ii, iii or iv	
					III	None required			
					IV	None required			

continued

Table 5.19 continued

Background		Suction	Key or bond	Drying shrinkage	Type of finish*	Suitable alternative undercoat plasters			Remarks
						Premixed lightweight plasters		Undercoat based on cement (for mix designation see Table 5.16)	
Class	Type					Gypsum	Cement		
	No-fines concrete	Low	Good	Low to moderate	I	–	–	ii, iii or iv	–
					II	Browning	Backing	ii, iii or iv	
					III	None required			
					IV	None required			
	Close textured or smooth lightweight aggregate concrete blocks	Low	Poor. Bonding treatment may be necessary	Moderate	II	Bonding	Backing or browning with bonding treatment	ii, iii or iv with bonding treatment	Background should be dry to minimize shrinkage. Sufficiently flat surfaces may be plastered with thin-coat plasters: bonding treatment may be necessary
					III	None required			
					IV	None required			
	Open textured lightweight aggregate concrete blocks	Low to moderate	Good	Moderate	I	–	–	iii or iv	Background should be dry to minimize shrinkage
					II	Browning	Backing or browning	iii or iv	
					III	None required			
					IV	None required			
	Autoclaved aerated concrete slabs and blockwork	Moderate to high	Moderate. Bonding treatment may be necessary	Moderate to high	I	–	–	iii or iv	High suction blocks may require special plasters of high water retentivity or sealing with a bonding agent. Sufficiently flat surfaces may be plastered with thin-coat plasters: bonding treatment may be necessary
					II	Browning or high suction browning	Backing or high suction backing	iii or iv	
					III	None required			
					IV				

	Background	Suction	Bond	Moisture movement	Type of finish	Undercoat	Finish*	Remarks
	Glazed bricks and tiled surfaces	Low	Poor. Bonding treatment is essential	Usually negligible	II	Bonding	—	Sufficiently flat surfaces may be plastered with thin-coat plasters. Plaster thickness should be kept to a minimum
					III	None required	Backing or browning	
					IV	None required	Backing or browning	
Slab	Wood wool cement slabs	Low	Good	Usually fixed dry, but may be high when used as permanent shuttering	II	Backing or browning	iv	—
Board	Plasterboard	Low	Adequate with suitable plasters	Fixed dry	I	Bonding	—	Sufficiently flat surfaces may be plastered with neat board finish plasters
					II	Bonding	—	
					III	None required	—	
					IV	None required	—	
	Expanded plastics boards	Low	Adequate with suitable plasters	Fixed dry	II	Bonding	—	Not suitable for paper-faced laminates
					III	None required	—	
					IV	None required	—	
Lathing	Metal lathing	Low	Good	None	I	—	ii and iii	Three coats required. Synthetic fibres may be incorporated in cement-based first undercoats
					II	Metal lathing	ii and iii	
					III	None required	—	Lathing to receive spray application available. Two coats may be necessary
					IV	None required	—	

* Types of finish (final coats):
I Gypsum   Neat gypsum finish plasters. A small proportion of lime, not exceeding 1 part of lime to 4 parts of gypsum plaster, may be added to certain finish plasters.
II Lightweight   Any of the proprietary finishes usually recommended for use on lightweight gypsum or cement plaster undercoats.
III Gypsum projection plasters   Gypsum projection plasters are spray applied in one coat which replaces the need for undercoat and finish plasters.
IV Gypsum one-coat plasters   Applied by hand in one coat which replaces the need for undercoat and finish plasters.
In addition, where good abrasion resistance is required, a cement-based finishing coat on a cement:sand or cement:lime:sand undercoat may be applied.

Source: Based on Table 4 BS 5492:1990

Table 5.20 Proprietary names for gypsum plasters

Type	Proprietary name
Plaster of Paris	British Gypsum superfine British Gypsum fine casting Crystacol Herculite
Retarded hemihydrate gypsum, plaster: Final coat plaster	Thistle finish plaster Sirapite B plaster
Board finish plaster	Thistle board finish plaster Diamond Universal Board Finish Plaster
Thin coat board finish plaster	Gyproc veneer finish Diamond Universal Board Finish Plaster
One-coat plaster	Snowplast patchwall plaster Snowplast universal plaster Thistle universal plaster Diamond universal one-coat plaster Diamond Universal 'X' One Coat Plaster
Premixed gypsum plaster	Thistle hardwall plaster

Premixed lightweight gypsum plaster:
  Undercoat plaster:
    browning plaster
    high suction browning plaster
    metal lathing plaster
    bonding plaster
	
	Carlite browning plaster
	Carlite browning HSB plaster
	Carlite metal lathing plaster
	Carlite bonding coat plaster

Final coat plaster	Carlite finish plaster Limelight finishing plaster
Renovation plasters: Undercoat plaster	 Snowplast renovating plaster system Thistle renovating undercoat plaster
Final coat plaster	Thistle renovating final coat plaster
X-ray resisting plaster: Undercoat plaster	 Barytite rough plaster Barytite fibred plaster
Final coat plaster	Barytite finish plaster
Projection plaster (machine applied)	Snowplast projection plaster Thistle projection plaster Lafarge PPM3 Hagalith L Sakrelith L Diamond Universal Projection Plaster

Note: Although these names are given in this standard, it is worth noting that this list is not exhaustive and the Gypsum Products Development Association has a wide range of member companies that are the major manufacturers and suppliers for the British Isles. The address for the Secretariate of the GPDA is: c/o KPMG, 165 Queen Victoria Street, London EC4 4DD.

Source: Based on Table 6 BS 5492:1990

of buildings. Sandstone is more resistant to this kind of direct chemical attack due to its pure silicate composition, but limestones and clays are vulnerable. Given the worst conditions, stone can decay as fast as steel.

All of these corrosive agents need a supply of water, which is of course plentiful in temperate climates. Corrosion can be reduced by minimizing the amount of water that lies on surfaces, or is retained in walling systems behind panels and in joints. Mortar in joints can be a source of soluble salts, producing efflorescence on stone faces which, by expanding, can apply pressure on rock crystals causing fracture. Any mechanical pressure causes fissures which open up the stone, exposing new and unseasoned areas to further deterioration. Soils, fungi and plants are also sources of decay. Architects in the 1950s and 1960s sometimes sprayed brick and stone walls in the country with diluted cow manure so that creepers and other life forms would take a hold to establish the picturesque. Weak acids from lichens roughen stone surfaces which then hold moisture better. Root systems from plant life will disrupt mortar, weaken stonework and accelerate decay. As a particular ecology is established bird droppings can be added to the list of corrosive agents speeding up disintegration. Finally, mechanical erosion by physical wear from this new community has an effect.

As a broad summary the chief minerals present in rocks are shown in Table 5.21. Greater detail is given later on particular types of stone.

As stone is not a homogeneous material and has a variety of pore structures, the uptake of water will vary. This is relevant for any treatment to stone: many treatments applied as solutions are only taken up in a patchy fashion. For this reason they should be avoided if at all possible.

The deterioration of stone can be minimized by laying the stone correctly: if a sedimentary stone is being used the natural grain should be horizontal, so that the load is taken perpendicular to the bedding planes. Rough surfaces collect water and sooty deposits, so it is better to choose smooth even polished surfaces for fast water run-off. Although dense stones wear better, the ability to resist freezing and spalling fracture may not necessarily be improved. (See section 4.5 on frost and freezing mechanisms.)

Recognition of stone types may be a problem, so the following short guide is included.

Guide to stone

Stone can first be classified according to its structure which will be a result of its geological origin and history. There are three main categories: *igneous, sedimentary* and *metamorphic,* All rocks are also a conglomerate of minerals so that if the minerals can be clearly identified, then a more precise classification can be given. This often requires looking at thin slices of rock under a microscope with polarized light.

Igneous rocks   These rocks are formed from molten material formed beneath the Earth's crust. When this material cools, different minerals are precipitated in a particular order, depending on the rate of cooling. The differing proportions of minerals so formed will determine the exact rock type. This process is known as *fractional crystallization*. The rate of cooling will also determine the size of the crystals. A fine-grained crystal structure (e.g. basalt) will indicate fast cooling, and a large-scale crystalline structure (e.g. granite) will show slow cooling, giving time for crystals to grow. Although there is a great variety of minerals in igneous rocks, they are chiefly silicates (salts of silicic acids). Igneous rocks have been classified according to the amount of silica present. If they have over 65% SiO_2 present, they are classified as *acidic*, between 52 and 65% SiO_2, they are *intermediate* and if less than 45% they are *ultrabasic*. Unfortunately, the concepts of 'acidic' and 'basic' used by geologists are not the same as those used by chemists measuring the pH of aqueous solutions. The broad division according to silica content has been kept by geologists, but the terms *silica, mafic* or *ultramafic* are now preferred.

Table 5.21 Rock types and their minerals

Rock type	Associated minerals	Density (kg/m³)
Granite	Feldspar, quartz micas	2600–2800
Slate	Biotite, muscovite, quartz feldspar	2700–2800
Sandstone	Quartz (with traces of mica)	2600–2650
	Feldspar (without quartz bonding)	2000–2650
Limestone	Calcite (dense)	2650–2850
	Calcite (other types)	1700–2600

Examples of igneous rocks used in building:

Granite (mottled white pink, grey and red). The mottled appearance is due to the large crystals which can be seen with the naked eye and which include silica (white), feldspar (cream or pink) and mica (black). There are also other minerals such as biotite, muscovite and hornblende. Granite has a relatively low density for a molten rock and contains large amounts of feldspar a mineral which has a low density and rises through the molten magma. Because of its concentration near the surface of the Earth it is an abundant mineral and crystallizes at a relatively low temperature.

Gabbro (grey, dark grey, black). The major minerals are bytownite and pyroxene, quartz, olivine and hornblende.

Serpentine (green, grey-green and black) is composed of olivine, pyroxene, hornblende, mica, garnet and iron oxide.

Sedimentary rocks As the Earth's crust weathers, major outcrops of rocks are worn down by wind and rain. Debris is carried down fast-flowing rivers. As the flow of the water decreases the size of the rocks that can be carried also becomes smaller. When rivers flow sluggishly into estuarine plains the particle size becomes very small, and silt beds and mud banks build up. This deposition continues until gradually, under its own weight, the layers of material become compacted. This whole process takes millions of years, but eventually the compacted material becomes hard, and when water recedes, the material forms one type of *sedimentary* rock. Subsequently, these deposits may be

compressed into hill or mountain formations and erode again. This whole process is sometimes referred to as the *rock cycle* (Fig. 5.20). On inspection the side of the rock is layered and this records the annual variation in the material laid down. (The laying down of layers of deposited material is known as *bedding* and the directions in which successive layers of bedding materials are laid are *bedding planes*. Within a single layer it may be possible to detect the deposition of larger grained material (laid down in winter and spring by heavier rivers), as against smaller grained material laid down in summer: this is *graded bedding*. If the stone is split it can reveal the fossilized action of wind creating ripples on an originally exposed sand. The whorls and patterns in some York *sandstones* are very clear. If the composition of the rock is studied closely with a hand lens, it is apparent that the individual particles are slightly rounded, showing the abrasive action of other stones when moved by wind, rain and water. Carbonate rocks form another major class of sedimentary rocks. They are created in shallow seas by the deposition of the skeletons of millions of microscopic sea creatures (*limestones*).

Sedimentary rocks can also be classified according to their age by the fossilized material present. As evolution progresses from fish to amphibia, reptiles and then mammals, individual fossilized remains can be identified and dated. This kind of classification is known as establishing a *stratigraphic column*.

Examples of sedimentary rocks used in building:

Limestone (white, grey, yellow and cream; red and brown varieties have iron impurities; black has organic components). These are calcareous

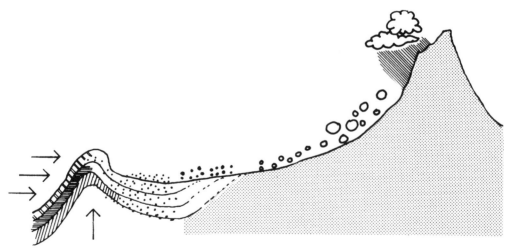

Figure 5.20 The rock cycle. Pressure from tectonic plate movement builds mountain features which are subsequently eroded by wind and rain, and fine grained particles eventually become compacted to form new rocks.

(calcium carbonate) rocks with fossil fragments that may be so extensive that the material is unsuitable for building. Oolitic *limestone* has calcareous grains (spherical structures) and other fossil fragments. *Travertine* is principally calcite produced from precipation, after water evaporation in caves and springs. Magnesium limestone is a general name for magnesium carbonate. *Chalk* is a pure variety of limestone, and can be white, cream or light grey.

Sandstone (brown, red, yellow, grey, white). These stones are known as *arenaceous* rocks (Latin for sand or quartz grained materials) with a grain size classified internationally as being above 0.2 mm and less than 2 mm in diameter. Sandstone is chiefly quartz, but if the grains are angular, the rock is often categorized as 'grit'. The grains of sandstone can be cemented by calcite or iron oxide.

Arkose (red, pink, grey) usually composed of more than 25% feldspar, the rest being quartz with some biotite and muscovite in a cementitious matrix of calcite and iron oxide. It is often derived from the erosion of early granitic material.

Greywacke (grey, black) has angular grains and is derived from coarse-grained quartz feldspar and large-scale rock fragments.

Metamorphic rocks   Metamorphic rocks have been subjected to great heat and pressure, which has consequently changed the state of the original material. Metamorphic rocks can often record particular signs of stress very clearly. The original (igneous) crystal structures in rocks can be deformed and flattened giving a characteristic directional pattern. Minerals can be subjected to such great heat that they re-crystallize and form entirely new mineral phases. The origin of this pressure is from large-scale crustal movements generated by tectonic plate movement. In the most active regions on Earth, around the Pacific, oceanic plates slip below continental plates forming destructive plate margins. There is enormous friction generated at these points and massive mountain ranges are built up, as in the mountain ranges on the East coasts of North and South America, with accompanying earthquake activity – for example on the San Andreas fault in California. Crustal movement and consequential compression where two blocks of continental crust collide was sufficient to cause deformation on a scale to produce the mountain building of the Himalayas. This process is known as *orogeny*.

The kind of metamorphic activity, associated with the building of mountain ranges, is known as *regional metamorphism*.

Examples are found in the metamorphosis of shales and mudstones to slates, and limestones to marble. The cleavage of slate is unusual. It does not relate to the bedding planes but occurs at right angles to the bedding planes. This is due to the re-crystallization of material which takes place with new crystal growth perpendicular to the pressure imposed and elongated parallel to the pressure zone. Hence the crystals are not isotropic but have directionality and are aligned.

Contact metamorphism is a different category of metamorphic activity: when hot igneous material intrudes up through the Earth's crust it affects adjacent rocks. Marble and spotted slate are common examples of this type of metamorphism.

Marble is a result of the re-crystallization of limestone to calcite. As the degree of metamorphosis increases relative to the amount of heat small features such as marine fossils are progressively altered until they are eliminated altogether. The original sedimentary bedding planes are often still distinguishable.

Examples of metamorphic rocks used in building:

Marble (from limestone: white, yellow, pink, red, green, black and strongly veined). Marble is chiefly composed of calcite with some dolomite and fossilized remains.

Quartzite (from quartz sandstone: white, red, grey). Interlocking quartz grains can be clearly seen.

Schist (from Greywacke (type of sandstone): whites, browns, red, grey). Schists are all formed by metamorphosis from former sedimentary structures and a particular schist is usually prefaced by the name of the chief mineral present. Characteristically all schists have some flaky minerals present, such as mica or chlorite, which can cause splitting.

Slates (from mudstones and shales giving a variety of colours from blacks to blues and greens, even light to dark browns, red and white). They are a group of very fine grained rocks. They exhibit perfect cleavage which is always perpendicular to the original bedding planes (due to the minerals re-orientating themselves after re-crystallization).

Identification   One of the easiest field tests for identifying rocks is to estimate their hardness, using Moh's scale of hardness and a small penknife (Table 4.2, section 4.2).

Selecting stone for building use

Within each category of stone, whether igneous, sedimentary or metamorphic, there are great variations in density and strength and, consequently, use. Within each category there is also enormous variation in colour. If it is necessary to try and establish the range of stone that might be available in the British Isles for building, there is an extensive range in the Geological Museum in London which can be checked for availability. The publication *Specification* also keeps an up-to-date list of stone types, their chief characteristics and possible sources. See also the *Natural Stone Directory* published by the Stone Federation (82, New Cavendish Street, London W1M 8AD. Tel 0171 580 5588) for up-to-date information about quarries.

Igneous types   Granites are the major building stones in this category and are easily recognizable with their large-scale interlocking crystal structure. They are commonly used for cobbles or setts, kerbstones, bollards, and often for the base stone courses in buildings. In large blocks (often weighing as much as ten tonnes) they are used as breakers for sea wall protection on eroded coasts, e.g. in Devon. When polished they make excellent cladding materials.

They deteriorate by the decay of individual minerals: black mica is often the most vulnerable mineral and can be softened by acid rain (the weak sulphuric acid in the atmosphere), decay or fall out leaving a pockmarked appearance (Fig. 5.21). Feldspar and ferro-magnesian silicate minerals can hydrate

(slowly absorb water) and over hundreds of years decay into clays. This is really a part of the process of erosion to sedimentary products.

Granites are available from Devon and Cornwall, Scotland in general, and the Channel Isles. We also import granite from Sweden, Norway and Portugal. Basalt is available from Scotland and Ireland, and gabbro from Scotland.

Sedimentary types   These soft rocks are the most commonly used in buildings. They include sandstone and limestone for walls, claddings, mouldings, etc. Because of their sedimentary nature with clear bedding planes, they can be easily split into sections. Some fine-grained limestones are so homogeneous in their structure that the bedding planes have to be indicated by *kerf marks* to avoid laying incorrectly. Some sedimentary stones can be split so thinly they have been used in the past as a traditional roofing material. Splitting these stones for such purposes used to be seasonal, relying on freeze–thaw cycles. Modern large scale refrigeration techniques may make it economic to once again make this traditional form of roofing material. Sandstones and limestones can be used for cladding panels but limestones are more likely to decay (Fig. 5.22). This is due to the fact that carbonate and sulphate minerals are more likely to dissolve as a result of weak acid corrosion. Chlorides from the sea can form weak hydrochloric acid and sulphur dioxide

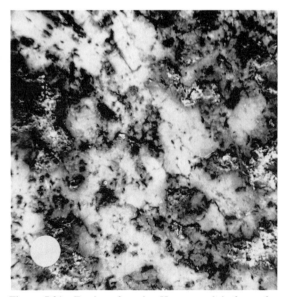

Figure 5.21   Erosion of granite. Horse trough in front of St Bartholomew's Hospital, London. This shows the erosion of individual crystals of mica leaving small cavities with surface.

Figure 5.22   St Pancras Station. The erosion of limestone showing deterioration of successive sedimentary layers. (Photograph taken in 1987.)

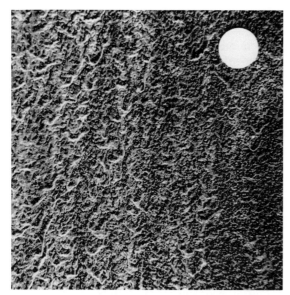

Figure 5.23 Camden Town Hall, London. The erosion of soft sedimentary stone by water movement. (Photograph taken in 1987.)

dissolved in rain water will form weak sulphuric acid. Sedimentary stones are more vulnerable to attack, not only because of their own chemical composition, but also because they have a cementitious matrix with minute pores. This degree of porosity is more likely to hold moisture and harmful salts (Fig. 5.23). The largest deposits of sandstone are found along the boundary of central southern England and Wales. Limestones are more widespread: Devonian beds (a grey colour range) are found in Southern England and extensive Carboniferous beds occur in the North of England with colour ranges from cream, to red, to light brown (in Durham, Derbyshire and Yorkshire). Around Yorkshire and Nottingham are the Permian limestones which come in rich yellow/brown and cream colours. In middle England are the Jurassic beds, from Lincolnshire, through Oxfordshire to Wiltshire with colours varying from cream to light brown and green/grey. In South-East England are the Cretaceous stones which are characteristically coarse-textured: from Kentish ragstone to the fine calcite structures in the chalks of Sussex.

Metamorphic types As these rocks have been subjected to heat and pressure they are normally stronger and more resistant to deterioration. Marbles and slates are the most common categories of metamorphic rocks in building use. Slates are used for roofing, flooring, table tops, etc. Slate is easily split

because its natural cleavage is always at right angles to its original bedding plane. Marbles are also used as flooring and as a decorative cladding material. Because of its homogeneity and ease of working, it is an ideal medium for carving sculpture. It can, however, decay almost as rapidly as its source material, limestone.

Marbles come from the most Northern parts of the British Isles: in Scotland and Ireland from the Isles of Skye and Iona, and County Galway. Carrera in Italy is still the largest marble-producing region in the world and the Italians have developed modern technologies for handling and finishing stone.

Sources for slate are the Lake District (green) and North Wales (grey, black and blue). There has been a decline in slate mining as seams have been worked out or become uneconomic. The waste heaps are an unfortunate reminder of a highly selective choice of building materials.

Quarrying The quarrying of stone requires a huge investment in machinery and labour. Quite often a large amount of overburden must be removed before it is possible to extract any usable stone. If there isn't an adequate natural system of breaks or 'joints', stone may have to be split by drilling and blasting. Freeze–thaw cycles can sometimes be used to split stone and, if it can be accelerated artificially under cover, annual and unpredictable cycles of freezing and thawing can be avoided. Most quarries are open, although slate is mined. The scale of underground workings has to be seen to be appreciated and those at Blaenau Ffestiniog are worth a visit. Once quarried, the stone is sawn with abrasives under a constant flow of water. Some sedimentary stones can be split if they show very distinct bedding planes, or cleaved in the case of slate. Polishing is highly mechanized, again with abrasives and water, although stone can be carved and dressed (surface-finished). It is worth remembering that apart from a plain finish, surfaces can also be 'punched', picked and tooled with fine ridges, or even 'margined' and 'vermiculated' or rusticated with different patterning.

Cast stone The only factor which distinguishes cast stone from concrete is that this product consciously simulates natural stone. The techniques for making it are very similar, although crushed natural stone (most often limestone) is used as aggregate, usually with white Portland cement with balancing pigments and selected sands. Cast stone can also be made from buff, light red and yellow sandstones, granite with mica or Portland stone. The crushed stone is usually the residue left after quarrying with aggregate sizes that do not exceed 25 mm.

Cast stone or 're-constructed stone' has been used for a long time as a substitute for real stone. The expense of making complex mouldings can be reduced by making moulds and then matching cast stone to natural stone to produce equivalent copies of elements in buildings. The use of cast stone is so widespread that it is difficult to tell the difference and many buildings dating from the early part of this century have used it. The reproduction of columns, mouldings, cills, balustrades, garden ornaments and artefacts (including temple follies) responds to a market demand for these elements at a reduced cost. Ornamental balls and finials are available in a range of sizes which are very useful for restoration. These items are available from a 'pattern book' just as, at the turn of the century, other elements were available in brick and terra cotta. The timescale for making these components is also relatively short and they can be of extremely high quality, depending on the aggregates selected and the control of strength in the mix.

There is also a growing market in the simulation of stone slates and blocks in new buildings. In restoration work it should not be used as a substitute for renovation using natural stone. However, cast stone is a high quality product, and careful specification will produce a material that is long-lasting and has wide application as facing panels, slabs and floor finishes. It is not just for the simulation or reproduction of architectural features. The current standard BS 1217 on cast stone is being rewritten, but it gives basic definitions, lists the aggregates used and details the production methods.

The mix design will vary from one manufacturer to another: different companies have individual 'recipes'. Depending on the natural stone aggregate used, the strength of the mix can be greater than a comparable ratio of cement, sand and aggregates in concrete. Although mixes of 1:3 are used (one part Portland cement to three parts aggregate), it is possible to reduce this to 1:4. The density of this mix gives a high-strength product with a minimum crushing strength of 25 N/mm^2. The mix can be altered to achieve crushing strengths up to 40 N/mm^2. Cast stone is less porous than stone, and is water- and frost-resistant. Unfortunately the hardness of the material makes it difficult to work by hand. However, as a cast material it can be moulded into most shapes and is ideally used when the production of a number of pieces make it viable to produce a mould. The production of a one-off shape is comparable to making the same shape from natural stone. The matching of existing pieces can be done by using moulds on site made of latex and fibreglass. They will reproduce all the fine detail of the original.

There are two main methods for casting. One is straightforward: as the wet cast method, it uses the same techniques as for making concrete. The second is the 'semi-dry' cast method: this limits water content, resulting in much higher strength, and uses compaction to achieve high density. This method can also allow a two-part method. The mould will be lined with the high-grade cast-stone mixture and a concrete mix will be compacted on top. Adhesion of the two materials is achieved through the compaction and simultaneous curing of both materials. This method is preferable to a two-layer product made using mechanical keying at the junction of both materials by the use of profiled moulds. Care has to be taken in the specification of aggregate to ensure conditions of purity and the elimination of materials that would be a source of decay or poor durability in concrete.

Cast or reconstituted stone can be repaired on site using a matching mix. It may be advisable to make up samples and allow them to harden to make sure there is a good colour match. Repair can only be made to components which are non-structural. The surface is prepared by cutting away friable material. Repair is similar to concrete: an initial spatterdash coat of 1:2 cement:sand of 3–5 mm thickness, with subsequent coats of 10 mm in thickness. One must allow for intermediate curing. As with any cement or concrete mixes the work should be protected from extremes of temperature and covered to prevent the evaporation of moisture. (See *Architects Journal*, 16.03.88, *Materials in refurbishment: stone and its repair*. Portland stone should always be replaced where possible by natural stone but small damaged areas can be make good by using a stone paste of white cement:lime:aggregate in the ratio of 1:2:8 where a mix of crushed Portland stone and silver sand is used instead of the aggregate.

Cast stone masonry units   These units are used for walling and the method of their manufacture is specified by BS 6457:1984. As they are structural units their constituents are more closely regulated in terms of the admixtures used. These have to conform to BS 5057 Parts 1 and 2, and aggregates and sand must have a chloride content below 0.2%. The block size is also limited and no dimension can exceed 650 mm. They are high-strength units with a compressive strength above 20 N/mm^2. The units should be clearly marked with the BS number, colour and finish.

Terrazzo   Although terrazzo is well known as a

decorative material it is still essentially a cast stone. It is used for flooring, staircases and also as vertical slabs for walling. Accessories can include profile skirting and cill details. On the continent it is also used for sinks, which may have an inset pattern of small tiles. The aggregates used in terrazzo are more exotic as colour is important, and the range of colours is usually achieved by using marble aggregates but other natural stones can be used with an aggregate range of 1 to 25 mm. Grey or white Portland cements can be used and it is normal to have a concrete background to minimize the use of the more expensive aggregates which form a layer either 12 mm thick (known as Type A) or 6 mm thick (Type B). The method of casting uses moulds in the same way as cast stone. The method of making hydraulically pressed terrazzo tiles is covered by BS 4131 and paving is covered by BS 4357.

The aggregates and cement are mixed dry and the water content is controlled to achieve maximum strength and to just permit high compaction. After the terrazzo units have been cast and cured any surface imperfections are filled using a grout or slurry. This is then ground off and the surface may also be polished for a greater reflectance giving brighter body colours to the aggregates. It is common for the cements used to be coloured with pigments or ground marble dust to match the body colour of the aggregate.

Fixings As the technology of fixing stone relies more and more on the use of metal accessories a successful stone cladding system depends on the construction of these thick veneer finishes, and not necessarily on the material itself. Stone varies considerably in its composition and the connection of the metal fixing into the stone itself is critical. The weight of a panel, if transferred completely to a small fixing point, can put a great deal of localized stress on the panel and initiate fracture. It is often thought that the problem is solved by increasing the number of fixing points to distribute the load. In fact the possibility of failure is not always diminished. As the number of fixing points increases, so does the number of components. As the system becomes more complex, a greater number of components can potentially fail. Sometimes stone is backed, e.g. with GRC to lighten the panelling system. This can simply introduce more problems, for instance, the nature of the adhesion between the two materials, and the longevity of the bond, which could degrade unpredictably over time.

Pollution The chief causes of pollution are rainwater saturated with carbon dioxide and sulphur dioxide. The weak acids then formed attack calcium carbonate: not just limestones, but also the calcareous deposits in sandstones. Further reaction between calcium carbonate and weak sulphuric acid solutions creates calcium sulphate which is more easily dissolved in rainwater, and crumbles away leaving new surfaces of stone open to environmental attack. Pollution is caused by small particles: from individual molecules to a maximum size of 20 microns. These airborne particles are *aerosols* (solid and liquid mixtures dissipated in air) and have great ability to penetrate into the body of porous material. Particles over 20 microns are likely to settle as deposits on surfaces.

The black crusts on the surface of buildings in London are usually a combination of organic compounds and sulphates (mostly calcium sulphate). This skin conceals an increasingly porous outer layer of stone. The skin does not form a protective layer as more advanced deterioration of stone is taking place underneath (Fig. 5.24). Small gypsum crystals formed initially become larger more expansive structures which put pressure on the new skin and can cause crumbling of the fabric. The black crusts are porous and act as a bridge for further contamination by gases

Figure 5.24 St Bartholomew's Hospital, London. Comparison between clean and uncleaned Portland stone. Note the shadow effects from blast cleaning. The cleaned surfaces have lost sharpness on their arrises. In cleaning stone, over-use of abrasives will remove the 'seasoned' skin which is slightly harder, and future decay can then happen at a faster rate. The deterioration of the existing base course emphasizes the need for a denser material in this position. (Photograph taken in 1987.)

and water vapour; they also retain dissolved gases and liquids which are no longer washed away.

This explains why a project to clean a building can turn into a project of substantial restoration. Before outlining a schedule of work, samples of stone should be taken from the facade to establish the real extent of damage by pollution.

Subflorescence Salts from ground water or impurities in mortars can be carried through the building fabric in a soluble state before crystallizing with a great increase in volume just below the surface. This is partly due to the evaporation of water from interior pores before it has had time to reach the surface. Different salts will also have different phase transition points and will solidify at different relative humidities. This internal crystallization puts internal stress on stone and can cause severe spalling. When the crystallization of salts takes place on the face of materials (where accelerated drying can take place), it is commonly known as efflorescence.

Organic acids Some lichens can exude weak organic acids that decay stone. Any kind of growth, of any size, from lichens or mosses, to plants or ivy can trap moisture causing damage in frosty conditions.

Feral pigeons Pigeons' droppings are corrosive. Facades should be designed so pigeons cannot live comfortably on ledges. They find difficulty in standing on surfaces raking at 70 degrees or more. Ledges can also be coated with a gel which feels unsafe to pigeons. All possible pigeon perches, such as pipes, the tops of brackets, ventilators and even overflow pipes should be removed. By limiting perches colonization can be discouraged. Natural predators such as trained hawks have been used as deterrents. Even stuffed predators such as owls can be quite a successful deterrent. If the fabric of the building cannot be altered, then spikes, wire, chicken mesh and polypropylene netting have to be used to prevent them alighting on possible perches. Pigeons also carry disease, through their droppings, although many pigeons in central London die from cancer rather than other causes. If people encourage pigeons by feeding them infestation can be rapid as they raise up to ten broods a year.

Pigeons can be culled as they are classed as a pest but they must be humanely destroyed by authorized persons and Local Authorities can take action to trap and destroy pigeons under the Wildlife and Countryside Act of 1981. More information can be obtained from the Royal Society for the Protection of Birds who recommend:

Depigeonal (Network Pest control services), telephone: 01925 75781
Pinnacle (Hughes and Hughes Ltd), telephone: 01402 349017

Racing pigeons are ringed and the Royal Pigeon Racing Association can trace their owners

Cleaning Cleaning is now a job with which an architect or designer might have to cope. We have seen that cleaning is not just for aesthetic reasons. Grime retains moisture and is made up largely of sooty particles which dissolve and initiate decay.

Cleaning can be carried out in a number of ways. If the surface is extensively wetted this may create even more problems, for example corrosion of remote metal fixings. Blast cleaning can use air, water or abrasives, or a combination. Care must be taken in specifying the particle size of abrasive to be used, and fine grades must be used for delicate cleaning. For guidance see BS 6270 Part 1:1982. Stone can be cleaned chemically but such solutions should always be used as a last resort and with specialist advice as they usually contain harsh alkalis or acids. Although these are used *in dilute solution* they can seriously damage facing material and should be used with great caution. All surfaces should be very carefully washed after application. Again, care must be taken in case there are adjacent metal fixings which could be vulnerable to electrochemical corrosion.

Introducing large amounts of water could also encourage latent efflorescence from trace salts present in the stone or, more likely, in any adjacent mortar. The whole of the building fabric should be carefully protected while cleaning is carried out, and glass must be protected from accidental 'sandblasting'. The following list gives basic cleaning methods.

Water	By spray, washing, steaming
Blast cleaning	Wet and dry abrasives
Chemical	Acids, acid salts, alkalis, organic solvents
Poulticing or dusting	Absorbency of contaminants and of aggressive cleaning agents after use.

Cleaning methods are compared in Table 5.22. Chemical cleaning agents are compared in Table 5.23.

Stabilization

After cleaning a surface it is very tempting to try to stabilize the surface and so extend the life of the finish

Table 5.22 Comparison of cleaning methods

Method	Process	Relative speed	Relative cost	Advantages	Disadvantages	Used on
Washing	Water spray	Slow	Low	No risk of damage to sound masonry except in frosty weather. No danger to operatives. Quiet	Limestone may develop brown patchy stains. Water penetration may damage interior finishes. Possible nuisance from spray and saturation of surrounding ground. Often requires supplementing with an abrasive method or high pressure water lance	Limestone Cast stone Brick
	Steam	Slow	High	Good for paint stripping (with chemicals)	As water spray but with less risk of moisture penetration. Not easy to obtain uniformly clean appearance	Any surface
	High pressure cold water	Medium	Medium	Useful where there are no complex profiles and water should be kept to minimum	As water spray	Limestone Cast stone Brick
	High pressure hot water	Medium	Medium	Improved softening of paint, grease, chewing gum. Improved thinning of viscous media	Needs great care in use. Danger of burns and scalds and damage to glazing and plastics fittings. Specialized pump required. Extra cost of fuel for heating	Any surface except marble and alabaster
	Poulticing, used after softening with mist sprays	Very slow	High	On vulnerable and valuable detail, considerable dirt may be removed by successive poultices without any abrasion	Slow method only suitable for high quality work	Polished marble
Abrasion	Dry air	Fast	High	No water to cause staining or internal damage. Can be used in any season	Risk of damage to surface being cleaned and to adjacent surfaces, including glass. Cannot be used on soft masonry. Possible noise and dust nuisance. Risk of drain blockage. Injurious dust from siliceous materials. For best results, needs to be followed by vigorous water washing. Can produce gun shading or mottled finish if operatives are unskilled	All except glazed brickwork and polished surfaces
	Wet air	Medium	High	Minimal amount of water. Less visible dust than with dry air abrasion. Uses less abrasive material than dry abrasion	Similar to dry air abrasion but greater risk of drain blockage. Some risk of staining limestone. Can result in mottled finish and gun shading if operatives are unskilled	All except glazed brickwork and polished surfaces

continued

Table 5.22 continued

Method	Process	Relative speed	Relative cost	Advantages	Disadvantages	Used on
	High pressure water	Medium/slow	High	Similar to high pressure water washing but with abrasive added. Less abrasive than wet air abrasion	Similar to dry air abrasion but greater risk of drain blockage. Some risk of staining limestone. Can result in mottled finish and gun shading if operatives are unskilled	All except glazed brickwork and polished surfaces
Mechanical	Carborundum disc	Slow	High	No water to cause staining or internal damage. Can be used in any season	Considerable risk of damage to surface, especially mouldings. Injurious dust from siliceous materials. Hand rubbing may be necessary for acceptable finish	Any surface
Chemical	Hydrofluoric acid (HF) based	Medium	Medium	Quiet. Will not damage painted surfaces	Needs extreme care in handling; can cause serious skin burns, and instant damage to unprotected glazing and polished surfaces. Scaffold pole ends needs to be plugged and boards carefully rinsed	Sandstone Unpolished granite Brick
	Hydrochloric acid (HCl) based	Medium	Medium	Quiet. Will not damage painted surfaces	Needs extreme care in handling. Any residue on stone will risk formation of soluble salts	Calcium carbonate stains
	Alkali based	Fast	Medium	Quiet. In exceptionally dirty conditions as a preliminary to other methods they may be used to soften deposits	Needs extreme care in use, can cause serious skin burns and damage to stained glass, metal surfaces and paint. Incorrect use can cause serious progressive damage to masonry. The material may lodge in joints and care should be taken to point open joints and to wash the stonework thoroughly after cleaning	Restricted areas Limestone
	Liquid detergent	Fast	Low	Quiet. Causes less damage than other chemical methods	May not remove heavier deposits	Glazed brickwork

Source: Based on Table 2 BS 6270:1982

Table 5.23 Chemical cleaning agents

Chemical	Use	Remarks
Liquid detergent	Removing material with organic binders. Cleaning clay brickwork or glazed brickwork	–
Alkali based agent (agent based on sodium hydroxide* (NaOH) or potassium hydroxide† (KOH))	Cleaning limestone	Should only be used in certain cases, e.g. removal of graffiti and softening up of heavy deposits prior to using other cleaning methods. Can leave deposits of harmful salts
Hydrochloric acid‡ (HCI) based agent	Removing cement splashes and calcium carbonate deposits. Cleaning newly erected brickwork	–
Hydrofluoric acid (HF) based agent	Cleaning sandstone, rough granite and clay and calcium silicate brickwork	Only chemical agent that does not leave soluble salts. However, long contact between it and any masonry surface and subsequent rinsing by water is likely to cause the deposition of white silica crystals in the surface pores. These crystals will be very difficult to eliminate. For safety reasons it is essential that these agents should be applied only by trained workers

* Commonly known as 'caustic soda'.
† Commonly known as 'caustic potash'.
‡ Commonly known as 'spirits of salts'.

Source: Based on Table 3 BS 6270:1982

that has just been achieved. It should be remembered that stone weathers naturally to some extent in the atmosphere and there is a certain degree of 'case hardening' on a fresh stone surface that could be degraded by harsh cleaning. Some treatments that have been applied to stone surfaces have proved harmful. As stone varies across its surface with respect to porosity, there will be an uneven distribution of any applied coating. At a later date treatments can appear blotchy and unsightly. Some silicon treatments attract dust particles electrostatically and this debris will be an unwelcome contamination. If in doubt over a particular specification it is worth checking with the manufacturer of the solution and inspecting stonework where it has already had several years weathering. Advice should also be taken from the BRE over particular systems and applications. The Society for the Protection of Ancient Buildings have also built up expertise on this subject and both owe a great deal to the expertise of John Ashurst, RIBA. (See *Cleaning Stone and Brick* by John Ashurst Technical pamphlet No. 4, published by the Society for the Protection of Ancient Buildings, 37 Spital Square, London E1.)

For a detailed study of all aspects of the durability of stone in the environment refer to *Stone Decay and Conservation* by Giovanni G. Amoroso, *Materials Science Monograph* No. 11, published by Elsevier,

1983. This book also deals with reaction from the different types of pollution and methods for protection and consolidation.

Treatments to stabilize stone surfaces include the following:

Organic treatments
    Acrylic resins
    Epoxy resins
    Silicones
    Vinyl polymers and co-polymers
    Waxes
    Unsaturated polyesters

It should be noted that all organic treatments will be subject to the same problems of environmental degradation discussed in section 2.4 on the degradation of thin surface films, and will have a limited life.

Inorganic treatments
    Alkali silicates
    Fluoro-silicates

Mortars   Repairs to mortar should be well specified. As usual, mortar strength should not exceed the compressive strength of the stone. The use of cement should be minimized, and all lime preparations should

Table 5.24 Mortar types and designations

Direction of change in properties is shown by the arrows	Mortar designation[1]	Type of mortar (proportion by volume)[2]					
		Cement:lime:sand[3,8]	Masonry cement:sand[4,8]	Cement:sand with plasticizer[4,8]	Hydraulic lime:sand[5,8]	Lime:pulverized fuel ash:sand[3,6,8]	Lime:brick dust:sand[3,7,8]
↑	i	$1:0$ to $\frac{1}{4}:3$	—	—	—	—	—
	ii	$1:\frac{1}{2}:4-4\frac{1}{2}$	$1:2\frac{1}{2}-3\frac{1}{2}$	$1:3-4$	—	—	—
	iii	$1:1:5-6$	$1:4-5$	$1:5-6$	—	—	—
Increasing ability to accommodate movement, e.g. due to settlement, temperature and moisture changes	iv	$1:2:8-9$	$1:5\frac{1}{2}-6\frac{1}{2}$	$1:7-8$	—	—	—
	v	$1:3:10-11$	$1:6\frac{1}{2}-7$	$1:8$	—	—	—
	vi	—	—	—	$2:5$	—	—
	vii	—	—	—	$1:3$	$2:1:5$	—
Increasing strength and durability	viii	—	—	—	—	—	$2:2:5$
	ix	$0:1:1$	—	—	—	$3:1:9$	—
	x	—	—	—	—	—	—
↓	xi	$0:2:5$	—	—	—	—	$1:1:3$

Notes:
1 Designations i–v correspond to the designations for mortar in other British Standards. Designations vi–xi are included for the purposes of this standard.
2 The proportions given are by volume and relate to lime putty and dry sand. The range of sand content is to allow for the effects of the differences in grading upon the properties of the mortar. Generally, the higher value is for sand that is well graded and the lower for coarse or uniformly fine sand. The designer should clearly indicate which proportions are required for the particular sand being used.
3 Unless otherwise stated, the term lime refers to non-hydraulic or semi-hydraulic lime. Lime may also be obtained in the form of lump quicklime.
4 The masonry cement mortars and plasticized mortars that are included in a given designation are of approximately equivalent strength to the corresponding cement:lime:sand mortars. These may be used at the discretion of the designer to improve workability and early frost resistance.
5 Hydraulic lime is lime which will set under water.
6 Pulverized fuel ash (PFA) with a low sulphate content assists the lime mortar to harden and enhances the hydraulic set.
7 Very finely ground clay brick reacts with lime to give a slight hydraulic set. The brick powder is generally red, brown or yellow in colour.
8 For the purposes of this table, the term sand includes natural sand, stone dust and other fine aggregates used to match the original material.

Source: Based on Table 4 BS 6270 Part 1:1982

be used with care as they are alkaline and can damage eyes and skin.

The specification of mortar and re-pointing application should be carefully controlled. There is a popular but unsatisfactory treatment to rubble walling in particular, where the mortar is finished proud. Where possible rubble walling should be finished flush pointed. Inset or proud mortar joints not only look terrible, but allow debris to collect. Masonry cement should comply with BS 5224, lime with BS 890 and sand generally with BS 1200.

Typical mixes (Source: *Architects' Journal Stone Handbook*)

Lime:sand	2:5
Cement:lime:sand	1:3:12
	1:2:9 (exposed details)
	1:1:6 (most sandstones)
	1:0:3 (dense granites only)

Mortar types and designations are listed in Table 5.24, while recommended mortar mixes for stonework are listed in Table 5.25.

BRE Digest No 362:June 1991 *Building Mortar* summarizes the use of mortar generally including control on site. Stone mortars that use masonry cement need higher ratios of cement and if high strength is needed for loadbearing masonry then greater control is needed, including air content for air-entraining mixtures.

General guidance A first source of reference should be BS 5390:1976 for advice on the use of stone, walling types, general work on site, restoration and cleaning. As stone forms, by whatever method, it will be placed under some pressure due to general crustal activity. Shrinkage cracks and settlement fissures develop. These natural breaks are sometimes referred to as *joints* and certainly ease the quarrying of material, but can limit the size of blocks to be quarried.

References To appreciate the use of stone as a building material and to quarry it sensitively with the minimum of waste requires geological knowledge. A general knowledge of the major stone-yielding areas in Britain is useful and where possible local resources should be used to save haulage costs.

The following publications may be useful:

Geological Survey Ten Mile Map published by the British Geological Survey (1979).
Natural Stone Directory, Ealing Publications Ltd, 73a High Street, Maidenhead, Berkshire.

Table 5.25 Recommended mortar mixes for stonework

Nature of stone	Internal walls	External walls			Paving		Strength and durability
		Sheltered exposure	Moderate exposure	Severe or marine exposure	Internal	External	
Highly durable	v	v	iv	iii	iii	iii	Increasing
e.g. basalt, granite,	vi	vi	vi	v	iv	iv	strength and
millstone grit	vii						durability
(for flint, see							
Note *)							
Moderately durable	v	v	iv	iii	iii	iii	Increasing
e.g. many limestones	vi	vi	v				strength and
and sandstones	vii	vii	vi	iv			durability
	ix	ix	vii	v	iv	iv	
Poorly durable	vii	vii	vi	v	v	v	Increasing
e.g. some calcareous	ix		vii	vi	vi	vi	strength and
sandstones, some	x						durability
fine pored	xi	ix	vii	vi	vi	vi	
limestones							

Note: In addition to the recommendations in this table, consideration should be given to the best local practice and to any mixes that have been developed to deal with special conditions. The numbers i–xi refer to mortar designations given in Table 5.24.
* Designation vi or vii should be used for repointing flint.

Source: Based on Table 5 BS 6270 Part 1:1982

Stone cladding

Stone claddings, pavements, kerbs and solid stone buildings form a very large part of the urban fabric. The qualities associated with stone, i.e. its durability and its large choice of colour and size, are also desirable in the interiors of buildings giving high quality and long lasting finishes. The initial cost is always offset in longer term by needing to replace the material and relatively low maintenance. Stone used in building cities often covers a huge range of geological examples. See *New Scientist* 23/30 December 1989, pp. 67–71, for an illustrated walk around London and also *London Illustrated Geological Walks* by Eric Robinson, Scottish Academic Press, 1984.

Principles for fixing cladding panels   Cladding which is remote from the building fabric, whether of precast concrete or stone, must be carried on fixings that are strong enough to support the calculated weight of the panel, and any other applied loads from wind or stress from building movement. Large panels are generally defined as exceeding 40 mm in thickness, and must be fixed back to the structure at every corner. Smaller stone fixings may only require two major structural fixings but should be fixed to adjacent tablets.

The whole building should be dimensionally stable. For steel-framed buildings, allowable deflection and degrees of rigidity should be *worked backwards* from the known tolerances and likely behaviour of the stone cladding panels. Concrete-framed buildings should have hardened fully so that complete shrinkage of the frame has taken place. It may be necessary to re-survey before designing the panels in detail. There should be adequate movement joints to allow for shrinkage which will not be complete even one year after casting structural concrete. Allow for approximately 12 mm of shrinkage for every 30 m length of concrete. Concrete can also expand if saturated with water by about 3 mm for every 6 m length. Compression joints of 13 mm should be incorporated on every floor and a sealant or a compressible polymer should be used. Vertical movement joints should also be designed at the corners of buildings: they should be at 1.5 to 3 mm from the corner where stress from thermal or moisture movements is most apparent.

Code of Practice 298:1972 has now been replaced by BS 8298:1989 *Design and Installation of Natural Stone Cladding*. It gives detailed advice on all aspects of using stone cladding. An important section on performance specification covers the main parameters that should be considered: site conditions, tolerances, types of stone, pointing, sealant and insulation, cavities, fixings and testing. It is quite common to build test rigs to check the performance of cladding for testing water penetration as well as checking the whole design. For a detailed checklist of 56 parameters that should be part of any performance specification, refer to BS 8200. This is a general standard that is applicable to all non-load-bearing vertical enclosures of buildings and cover all aspects of strength, water penetration and thermal performance.

Fixings types: load-bearing and restraining   Stone or precast panels are always rebated. The panels hang either on nibs or ledges, to which they are bolted, or simply on upward-pointing metal cramps which are set securely into and cantilever from the main structure. A direct fixing through the panel is also acceptable and can be exposed (and allow for water shedding) or concealed with holes filled with stone pellets. Additional fixing positions which do not take the full structural load of the panel but which steady the panel against additional wind loadings or thermal stresses are called *restraint fixings*. These restraint fixings are shaped like a straightforward metal tie or downward-pointing cramp. All metal fixings whether load-bearing or restraint should be mechanically dovetailed into the backing construction to avoid being pulled out.

A high degree of accuracy is required in setting out the dimensioning for fixings. It should not be less than Grade 1 in PD 6440 and preferably 'Special Grade'.

The type of fixings should be chosen and detailed in conjunction with a fixings specialist, preferably with the manufacturer of the fixings who can then advise on structural strength, exact alloy specification, compatibility with other fixings in terms of electrochemical or 'bi-metallic' corrosion and atmospheric corrosion generally. The metals normally used are: stainless steel, copper-based alloys which include aluminium bronze, silicon-aluminium bronze, phosphor bronze and copper. Unfortunately, copper can stain water run-off. Many metals react with each other electrochemically and must be kept apart. Non-metallic washers and brushes should always be used as separators. See Table 5.26 for details of alloys used for metal components in masonry.

Panel thicknesses   The thickness of the panel will depend on the type of stone used. Panels of sedimentary stones should be fairly thick, e.g. 50–100 mm depending on the strength of the particular stone. The minimum thickness is normally 75 mm to allow for variations in the type of stone available. Marble and granites can be thinner, between 20 and

Table 5.26 Materials for metal components in masonry

Base material	Form	Grade and standard to be complied with	Protective measures carried out after fabrication
Copper	–	BS 6017	–
Copper alloys	–	BS 2870:1980, grades listed in Tables 8 and 12	Material other than phosphor bronze to be formed either:
		BS 2873:1969, grades listed in Tables 4 and 6	(a) by bending at dull red heat and allowing to cool in still air, or
		BS 2874:1969, grades listed in Tables 6, 8 and 9 *except* CA 106	(b) by cold forming and subsequently stress relief annealing at 250 °C to 300 °C–60 minutes
			Effectiveness of stress relieving of cold formed components to be tested by the supplier using the mercurous nitrate test described in clause 11 of BS 2874:1969
Austenitic stainless steel, minimum 18/8 composition and excluding free machining specifications	Strip	BS 1449:Part 2	–
	Bar/rod	BS 970:Part 1	–
	Tube	BS 6323:Part 8	–
	Wire	BS 1554 BS 3111:Part 2	–

Source: Based on Table 6 BS 6270 Part 1:1982

40 mm depending on their situation. Slates should be at least 40 mm thick wherever they are used.

In general, whatever the thickness of stone, fixings should not go deeper than *halfway* into the stone. This ratio can be less for granite (which is harder) when used in situations below first-floor level, but should be more for marble used in exposed conditions above first-floor level.

Mortar and joints Mortar should not be stronger than the stone. It is usually made up from ground dust of the stone to be used. The maximum width of joint should be 13 mm. Joints should be pre-wetted before applying mortar. Recommended minimum mortar joint widths and mortar mixtures are shown in Table 5.27.

Sometimes dry joints may be used. Panels should be designed not to abut (which will not cater for any movement), not to be close (which will encourage water penetration by capillary action) and, if possible, incorporate a rebate which will give some weather protection. Any horizontal rebates should be sloped to allow for drainage.

Sealants There is a variety of sealants on the market. They are usually applied by gun, sometimes with a knife, and include the following compounds:

Acrylic resin (15 year life)
One and two-part polysulphide (20 year life)
One and two-part polyurethane
Silicone-based compounds

Joint widths will vary according to which sealant is used, but are generally between 13 and 25 mm. Their depth should not be less than 7 mm. A backing material should be provided for sealants. Expanded rubber or polyethylene are compatible with the above sealants but should not be located in such a way that cavities will be bridged. Some sealants can now last for 20 years and they can be stripped out and renewed easily.

Table 5.27 Minimum mortar joint widths

Stone type	Mortar mix	Joint width (mm)
Limestone and sandstone	1:5:7 to 1:2:8	5
Granite	1:4	5
Slate	1:4	3
Riven slate	1:4	7
Marble	1:4	3

Recent failures in cladding and floor tiles are shown by discolouration around the joints in wide grey bands. A very recent example is the Vermont marble at 20 Cabot Square, Canary Wharf. This is thought to be due to the application of the wrong primer before the silicone sealant was applied or the omission of a primer. The oil from the sealant was taken up by the stone and discolouration varies according to the stone density. The lowest risk sealants are acrylic and polysulphide types. Oil-based mastics and butyl mastics cause problems during application. Poly-urethane sealants need a primer and silicone sealants are the most likely to stain. The Stone Federation offers guidance in their leaflet *Considerations to reduce staining on natural stone facades*.

Movement   There must be adequate allowance for movement in three dimensions when setting out on site. This is difficult to achieve with individual fixing points, and the use of channel systems makes it easier to adjust panels. When a building is insulated externally one has to use more remote fixings which cantilever the slab further out from the main structure. This puts far more load on the fixings and, in turn, stress on the individual panel, as the calculable force will be a function of the load of the cladding and its distance from the main structure.

Vertical movement joints should be at least every 6 m and horizontally every 12 m. These joints will have to cater for possible contraction of a concrete structural frame or cementitious materials as well as expansion.

Fixings: metals   All fixings should be non-ferrous except for stainless steel. Differing metals should not be fixed to each other or adjacent to each other. If this is unavoidable they must be separated from each other with a non-conducting medium such as PTFE (polytetrafluoroethylene) as washers or separating layers. Metal fixings described in Code of Practice 298:1972 are as follows:

Load-bearing: aluminium   bronze,   silicon-aluminium bronze, stainless steel.
Tying-back fixings: copper, phosphor bronze, silicon-aluminium bronze, stainless steel. See CP 297 and 298 for non-ferrous fixings.

Substrate structure   Load-bearing fixings are most effective if they are set into concrete. Fixings set into brickwork require careful spacing to hit the body of a brick and not the edge detail, or should be securely coursed in. The brickwork should be detailed as solid, otherwise great care has to be taken with fixings. Fixings into blockwork (especially lightweight) will probably be unsatisfactory.

Cavities   It is becoming more common to set the stone cladding away from the background for these reasons.

- To ensure that any soluble salts present in backing materials and mortar are not able to crystallize and cause damage to the stone or its bedding.
- To eliminate the effect of water penetration by capillary action between the stone and its background.
- To allow for insulation.

Cavities are recommended to be a minimum of 20 mm deep but should not exceed 100 mm because of the additional strain on fixings. Wider cavities will need separate framing structures rather than relying on individual fixings.

Pre-cast cladding and stone   An alternative method of construction is to use pre-cast concrete panels with a stone veneer. The stone thicknesses will be the same but their mounting is different. The stone should be fixed by stainless steel dowels (or equivalent strength non-ferrous material) in the concrete and by epoxy resins back to the stone. There should be rubber grommets and a separating layer between the concrete and the stone. This gives a large scale structural walling component that is ready to receive glazing. As a composite piece of construction, it should be well detailed and insulated to avoid cold bridges. The method saves time on site and gives a high measure of quality control in the fabrication of the reinforced panels. Large and complex sections can be made: up to 24 metres long (6 stories). As individual panels can be many tonnes in weight, site handling should be well organized and panels should be craned from delivery straight to fixing, and not stored. This is to avoid double handling and any damage on site. There are long delivery times for such large composite constructions and six months at least should be allowed.

Learning by case studies   It is better to learn techniques from current practice and to relate different approaches to basic principles. Quite often a decision has to be changed because of unexpected problems or a change in timescale, and the repercussions should be thought out first. The Architectural Press publish Construction Studies in the *Architects Journal*. These are invaluable to the student and practitioner. For example, in the issue of 13th May 1983 Wilfried Dechau discusses the Staatsgalerie by Stirling and Wilford. The gallery is clad with stone panels fixed on stainless steel brackets with open joints, a 40 mm cavity and 60 mm mineral wool insulation. Originally the whole building was to be clad with sandstone but the size of some of the panels was so great that their thickness would have had to be increased to 60 mm from 40 mm. This was unacceptable, so travertine was substituted.

There are some unusual features to this stone cladding. First the mixing of stone could be problematic,[2] and secondly, the stone is cut not only *with* the bed but also *across* it. Weathering will intensify the differences and give a pronounced texture, which could be interesting although it could make the building more vulnerable to damage in freezing conditions.

One of the great problems in having cantilevered stone panelling is the joints. Mortar seems unsatisfactory as it gives a false sense of a continuous wall. Stirling and Wilford have used open joints which is the honest solution. The shadow gap conceals the insulation behind. There are problems with driving rain, particularly with continuous joints around major openings, and so a different technology to deal with water run-off has had to be incorporated.

For detailed advice on tolerances, finishing and workmanship, reference should be made to BS 8298. This code of practice should not be the only advice sought. Stone cladding cannot be understood well if studied in isolation. The choice of stone, design of fixings, and final jointing methods are very much a team effort between designer, stone supplier, fixings manufacturer and subcontractor, and all must be involved early on in a contract as their advice will affect how fixings are cast into reinforced concrete, and to what tolerances an overall structure will be built. Latest BRE publications highlight poor practice and give practical guidance.

5.6 Flooring

All floorings in this category are hard, cold and noisy. A major advantage is the ability to provide some thermal storage particularly if exposed to daylight for passive solar energy gains. These are traditionally long-lasting finishes but the preparation of the substrate and laying of the floor should be undertaken with care to avoid cracking.

Stone flooring   Stone is obviously a highly durable solution for flooring and can be used for a range of buildings, whether domestic or commercial. All stone floors are slippery when wet apart from sandstone. Water and mild detergents should be used for cleaning and polishes should be avoided as they can make walking surfaces slippery and dangerous. For sources of stone and description of types, see *Selecting stone for building use* in this section. See also BS 8000 *Workmanship on building sites*, section 11.2 on natural stone tiling. Granites and slates are both completely resistant to water, hence their use as damp-proof courses. Slate floors were commonly used in Wales, usually laid as large slabs. Sandstones and limestones vary greatly in density and the less dense stones are more absorbent and more likely to be stained by oils and grease. Slate can also stain and, although there is a tendency to use seals to prevent this, it is not recommended. See BS 5385 Part 3:1988.

The more abundant sedimentary stones are the least expensive. Slate and marble are often double the price. Granites can be equivalent in price to the metamorphic stones and also much higher depending on colour and availability.

Reconstituted stone can also be used for flooring systems. The methods of bedding are similar and so is the advice for cleaning.

Concrete finishes   These are traditionally cheap and more suited to industrial use. But there is great potential in the choice of aggregates for producing special effects. Nigel Coates embedded a variety of objects in the floor of Cafe Bongo in Tokyo to great effect with a resin coating applied afterwards. The concrete is resistant to most alkalis and acids but a surface texture is needed if the finish is not to be slippery when wet. Curing will take at least 7 days. After curing, one must wait before applying floor finishes. Slabs of 150 mm with 50 mm screed toppings can take over a year to harden thoroughly. Floating floors on top of insulation are becoming more common as thermal standard in construction improve but, to avoid failure, these toppings should be at least 75 mm in thickness and the British Standard 8203 should be followed. Mesh reinforcement can produce its own problems: heavier grades prevent compaction of screed material. Power floating will provide a good smooth industrial flooring finish and remove the necessity for a finishing screed but is more expensive. Additional costs may be justified by not having to wait until the screed has fully hardened and thus speeding up the time for handover.

Terrazzo flooring   As with stone flooring, neutral soap solutions are advisable for cleaning followed by

Table 5.28 Ceramic flooring applications

Type of ceramic	Method of laying	Sizes and indicative price per square metre (1994)
Stone		
Slate: dark grey, green, mottled	Bed on 1:1:3 mix with concrete base. Slurry on back	1500 × 900 × 40 mm Price: £75–138
Marble: travertine	Bed on 1:1:3 mix Slurry on back	1500 × 900 × 15–30 mm Price: £160–200
Granites: crystalline with a mottled appearance; grey pink, black, white; granite and whinstone (BS 435:1993) kerbs, channels, quadrants and setts	Bed on 1:1:3 mix with concrete base	1830 × 910 × 40 mm Price: £200–375 (tiles) Price: £43–48 (setts)
Quartzite: crystalline, whitish grey	Bed on 1:1:3 mix Slurry on back	152–229 × 10–20 mm Tiles/rectangular shapes up to 900 mm long
Sandstones: brown, red, yellow, grey, white	Bed on 1:1:5–6 mix on concrete base Slurry on back Sand bedding of 25 mm for external use	3050 × 1220 × 40–75 mm Price: £90–130 Price: £55–60 (York stone slabs)
Limestones: white, cream, grey	Bed on 1:1:5–6 mix on concrete base Slurry on back	1500 × 900 × 15–50 mm Price: £110–180
Concrete		
Concrete *in situ* Portland cement to BS 12:1971 Aggregates to BS 1198–1200:1976 Colour dependent on cement and aggregates used. Texture depending on method of compaction/power floating after placing BS 8203 and 8204:1988 Section 2 *In situ floor finishes*	On site pouring mechanically or by hand Finishing by hand trowelling or grinding or power floating Surface quality can be altered by embedding a top layer of aggregates or other materials	Laid in maximum recommended widths of 4500 mm; for surface control, contraction joints to be not more than 6000 mm apart Price: £38–62
Concrete tiles and slabs Colour dependent on cement and aggregates used; fine and deep textures can be case BS 1217:1986 *Cast stone* BS 1197:1980 *Concrete flooring, tiles and fittings* BS 368:1971 *Pre-cast concrete flags*	Flags and paving should be bedded on sand on graded hardcore *Note*: Sizes given are nominal, work sizes are 2 mm less	Flagstones: 450/600/750/900 × 600 mm (length × width), 50–70 mm thick; or 300/400/450 mm square, 50 or 63 mm thick Precast concrete tiles: 150 × 150 × 15 to 500 × 500 × 40 mm, with a good range in between; or 200/300/400 mm square Price: £8–15
Terrazzo		
Terrazzo *in situ* Marble aggregate set in white/tinted Portland cement, ground and polished finish to expose aggregate of >6 mm size BS 8204 Section 3	Laid as screeds but compacted and set in bays ready for mechanical finishing	Lay in bays with dividing bronze or stainless strips to specialist contractor's recommendations Take care to align with structural joints in screeds and slabs. Thicknesses are typically 12 mm on structural concrete or 15 mm on screeds. Price: £58–75

continued

Table 5.28 continued

Type of ceramic	Method of laying	Sizes and indicative price per square metre (1994)
Terrazzo tiles As *in situ* terrazzo but with concrete backing BS 4131:1973 *Terrazzo tiles* BS 4357:1968 *Precast terrazzo units*	See BS 5385:1988 *Bedding in sand/cement mortar on concrete base* Tile backs coated with slurry before placing or receiving adhesive	$150 \times 150 \times 15$ to $500 \times 500 \times 40$ mm Price: £58–75
Bricks Bricks: from clay bodies pressed and fired; great range of colours and density with variation in laying BS 3921:1985	Either bedded on sand or in weak mortar on graded hardcore for foot traffic, or bedded in stronger mortar on concrete bases for heavier traffic	Standard size: $215 \times 102.5 \times 65$ mm Price: £33–50
Brick paviours: greater range of textures with chequers, ribs etc.	As above; for internal use bed on screed	$190 \times 95 \times 19$, $216 \times 108 \times 19$, $214 \times 114 \times 51$ or 251×124 51 mm Price: £29–35
Tiles Quarries Clay bodies pressed and fired with colours from natural clays: red, buff, blue, brown; range of accessories with cills, stairtreads, special coves, skirtings BS 6431 Part 1:1983, Part 2:1984, Parts 3, 4 & 5:1986	Laid with 15–20 mm mortar on screed or concrete, or cement-based adhesive to manufacturer's instructions Damp-proof membranes may be needed to prevent salts rising through joints (efflorescence)	Modular size: 100 or 200 mm square, 19 mm thickness See Manufacturers' literature for greater range of sizes, generally 152–229 mm square but with greatly increased depth of 32 mm for the larger size. Price: £19–37
Ceramic floor tiles BS 6431 Part 1:1983 The greatest variety of colour and finish in pressed and extruded tiles from refined clays with a range of glazes and varying density and frost resistance dependent on firing and degree of vitrification; ribbed, chequered and non-slip carbide treated surfaces	Mortar- or cement-based adhesives on screeds, and some cases plywood if the backing is rigid enough. Manufacturers' instructions have to be followed to prevent failure	See co-ordinating sizes in Section 5.10 *Tile dimensioning* Modular co-ordinating sizes are: $100 \times 100 \times 9.5$ mm or $200 \times 200 \times 9.5$ mm, but sizes range from 50×50 to 300×300 mm Price: £30–80
Mosaic tiles or tesserae Small ceramic or glass shapes usually bonded to a light mesh and supplied often with facing paper so they can be laid in larger areas; great range of colours; used in pools, bathrooms as greater number of joints give resistance to slipping; effective as external cladding. BS 5385 Parts 3 & 4:1986	Same process as for ceramic tiles After setting the facing paper is removed and joints grouted	5–13 mm squares, rectangles, hexagons and round shapes Price: £45–60

washing with warm water. Polishes and seals should be avoided as they are slippery. Fine abrasives can be used to remove stains and bring up the surface generally.

Magnesium oxychloride   This material is made by taking the mineral *magnesite* (magnesium carbonate) which comes in natural finishes that are colourless, white or greyish white and heating to high temperature to remove carbon dioxide and water, giving magnesium oxide. When combined with magnesium chloride a strong hard-setting cement called *magnesium oxychloride* is formed. (Note: Oxychloride cements can in theory also be formed with other oxides). One problem is that chloride ions can be released from this material, especially if the surface is wetted. Chloride ions attack metals causing corrosion. Because of this and the possibility of staining from corrosion deposits, the flooring should be removed from contact with metals and special care should be taken with cleaning and floor sealants.

Table 5.28 gives information about ceramic flooring applications.

Notes

1 Walter Gropius, Bauhaus books, V. 12. The quotation is shown adjacent to a house (1926) with a completely tiled kitchen.

2   Advice on the juxtaposition of different stones is difficult to obtain. Concern that used to be expressed over putting other types of stone below sandstones may be diminishing, given a generally harsher environment in which decay happens at a more rapid rate, and other factors which are outside the control of an architect. Quartz in sandstone dissolves very slightly over time to give the weak silicic acid, H_4SiO_4, and this will have an effect on adjacent limestones in a long-life building. Although stone is always regarded as a long-life material the rate of decay can in fact be comparable with that of rusting steel if local environmental conditions are poor. The continual re-furbishment of soft sedimentary stones in inner cities has become a reality.

6 Metals

6.1 Introduction

Metals are the group of materials that are at most risk in our atmosphere. In their natural state they exist as oxides and sulphides before extraction into pure elements. Once extracted from their ore they can be highly reactive in the atmosphere and can recombine with oxygen and, in some cases, sulphur dioxide, to form more stable compounds, like those that they originated from. Metals come in an order of reactivity. Those metals that are found in a pure state in the Earth's crust, such as silver or gold, mercury or copper, are known as 'noble' metals. Those metals that are more reactive are known as 'base' metals, and include all our more commonly used building materials such as lead, iron and aluminium. *Base* and *noble* metals fit into a classification of reactivity, the electrochemical series, which rates the *electrode potential* of every metal. Electrode potential is a physical measurement which describes the state of equilibrium between a metal and its ions, as follows:

$$M^{n+}_{(aq)} + ne \rightleftharpoons M_{(s)}$$

Equilibrium for base metals (Tend to dissolve with a loss of electrons)

Equilibrium for noble metals (Tend to stay stable with no loss of electrons)

Table 6.1 gives values of electrode potentials for a variety of metals.

6.2 Corrosion

As the metals used in the building industry all have a tendency to corrode, especially steel as an iron alloy (Fig. 6.1), our efforts to keep them stable rely on the effectiveness of coatings and finishes. Alternatively,

we can specify alloys that are stable, often by their ability to form tenacious oxide coatings that are difficult for water and air to penetrate. One example of this is the alloying of chromium with steel to make stainless steel. This alloy works by selective oxidation: the chromium oxidizes first forming a chromium oxide coating which protects the metal below. There should be at least 13% chromium in the alloy for effective protection. Chromium steel alloys have replaced the earlier technique of chromium-plating steel. If the chromium coating was damaged then the steel below would still corrode.

Coatings applied to metals must do the following:

- Protect the metal substrate from initiators of corrosion such as air or water, and actively inhibit corrosion.
- Provide effective bonding to the metal.
- Be compatible with the metal so that thermal movement can be undertaken by metal and coating together.

Table 6.1 Electrochemical series: standard electrode potentials for half-cell reactions

Metal	Half reaction	$E^{\oplus}$/V
Gold	$Au^{3+} + 3e^- \rightarrow Au$	1.42
Silver	$Ag^+ + e^- \rightarrow Ag$	0.80
Copper	$Cu^{2+} + 2e^- \rightarrow Cu$	0.34
Hydrogen (reference)	$H^+ + e^- \rightarrow \frac{1}{2}H_2$	0.00
Lead	$Pb^{2+} + 2e^- \rightarrow Pb$	−0.13
Iron	$Fe^{2+} + 2e^- \rightarrow Fe$	−0.44
Zinc	$Zn^{2+} + 2e^- \rightarrow Zn$	−0.76
Aluminium	$Al^{3+} + 3e^- \rightarrow Al$	−1.66

Figure 6.1   Complete decay of steel. The metal of this steel wheel barrow is reduced to a paper thin state and the weight of an adjacent section is sufficient to cause fracture. Once steel is exposed and starts to corrode, the conversion of steel into rust (hydrated iron oxide) is progressive and continual until all the metal is converted. As the rust deposit forms, it does not provide a layer that can protect the unaffected inner metal; it is so porous that air and water vapour can diffuse through and initiate further conversion.

Coating systems

The most important part of coating a metal is the initial cleaning and preparation. As the metals most commonly coated are irons and steels, the purpose of cleaning is generally to remove an oxide film (or millscale if the section has been rolled). See Code of Practice 3012:1972 *Code for cleaning and preparation of metal surfaces*. The quality of surface finish is important and blast-cleaned surfaces have to be specified. See BS 7079 *Preparation of steel substrates before application of paint and related products*. This standard is now in 6 parts: Part 0:1990 gives an introduction and glossary, Group A (Parts A1–A2) give a visual assessment of surface cleanliness, specifying rust grades and preparation grades for uncoated steel with photographs, Group B (Parts B1–B3) give methods to assess cleanliness and Group C (Parts C1–C4) give the surface roughness characteristics of blast-cleaned steel substrates.

Swedish standards used to be quoted for this work. They are now incorporated into International Standard ISO 850–1:1988 where grades for surface preparation are given after cleaning. Grade Sa 2.5 is generally acceptable and Sa 3 is quoted for surfaces requiring above-average protection. For steels needing a metal finish cast iron grit is recommended to provide a bright

Figure 6.2   Port buildings, Manhattan Island. These metal buildings rely heavily on paint films for protection. Their construction and maintenance owe a lot to the technology of bridge building.

surface with an idealized angular 'peaked' surface profile for good adhesion. There are two main methods of blast cleaning (or 'mechanical blasting'): using steel shot which gives a rounded profile to the surface topography, and grit blasting which gives a more angular profile. One drawback of mechanical blasting (especially using grit) is that contaminants can be left behind and metal surfaces should then ideally be washed. This is best carried out under factory conditions since for best results the metal should be heated to 500 °C for at least 90 minutes. Angular topography gives a greater surface area of exposure. The higher topography requires a greater thickness of applied coating system, which should be three times average maximum profile (AMP), according to ISO standards (8503 Parts 1–4).

Coating systems can be organic or inorganic, but are preceded by phosphate treatments which act as corrosion inhibitors. Wherever possible these finishes should be carried out in controlled conditions, particularly to enable blast-cleaning to be properly carried out. This is not always possible on site and the very minimum standards for site preparation should allow for cleaning with power tools (wire brushing is not adequate). See *Painting Steelwork* by I.P. Haigh, Construction Industry Research and Information Association Report 93, 1982. British Steel produces a series of leaflets on corrosion protection giving guidance for surface preparation and coating systems depending on location. Steel buildings can of course be successfully protected in the long term (Fig. 6.2). There are many other uses for galvanized steel (Fig. 6.3).

Galvanizing   Galvanizing is the process of dipping steel into molten zinc. The steel is cleaned and then dipped into hydrochloric acid, which produces iron chloride on the surface of the metal which acts as a flux. When the cleaned steel is dipped into molten zinc at 450 °C, new inter-metallic layers are formed between the zinc and the steel. These compounds are relatively brittle (Fig. 6.3). See BS 729:1986 *Specification for hot–dip galvanized coatings on iron and steel*. Galvanizing works because zinc is anodic to steel: if the coating is scratched then the zinc will decay preferentially to the steel and so offer protection.

Galvanizing is used where steel needs a coating which will have a guaranteed life of over 20 years. This is in the worst case of poor environmental conditions; for example in the city, as most coatings will survive for over 30 years in rural conditions. Coating thickness varies from 50 microns on thin steel to 100 microns on thick steel. Coating specifications are given by weight for example, $100\,\mathrm{g/m^2} = 14$

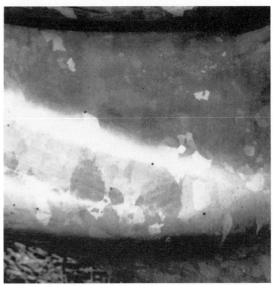

Figure 6.3   Galvanized crash barrier, Northern Spain. Giant crystals of zinc have grown during the cooling of the zinc layer on the underlying steel.

micron thickness of coating. For structural steel (over 5 mm in thickness) the weight will be $610\,\mathrm{g/m^2}$. A minimum coating would be 85 microns and for a long life 140–210 microns would be preferred. The higher end of this recommendation would be for sea splash zones. Coatings can be repaired with zinc-rich paints. See also International Standard ISO 146 and *Galvanizing for structural steelwork* by the *Zinc Development Association* (1990).

Figure 6.4   Empire State Building, Manhattan. The use of gold can be justified as a finish for lettering in a very inaccessible place.

Steel can be weakened by heating during galvanizing. After heating up to 450 °C, metal previously stressed through coldworking or welding can be stress-relieved and there can be some local distortion and weakening around pre-worked areas.

Phosphoric acid is used to pre-treat steel and give a micro-roughening to the metal surface. Paint systems that adhere well to galvanized steel are based on calcium plumbate, chlorinated rubber or epoxy polyamide.

Aluminium and tin coatings on steel   Steel can also be dipped into molten tin but this is an expensive process and sometimes dipping into an alloy of lead (88–75%) and tin is used for engineering parts in cars. Steel can also be coated with aluminium by dipping in the molten metal, after rigorous cleaning, giving a coating thickness of 50–75 microns.

Sheradizing   This process of applying a zinc coating is different from galvanizing; instead of dipping in molten metal, the items to be treated are rotated in a drum containing powdered metal zinc which, at high temperatures, diffuses into the surface of the hot metal items. This gives a harder outer coating than

Figure 6.5 (right)   Entrance to the McGraw Hill building, New York (Hood, Godley and Fowlhoux, 1930–31). Blue enamelled panels and nickel-rich alloys have ensured that this entrance has survived as originally detailed for over 60 years.

Figure 6.6   Chrysler Building, New York (William van Allen, 1930). In 1930 the McGraw Hill building was the largest to use terra cotta as a facing material. It had a glazed green finish. These two entrances are typical of the fashion in America during this period: entrances had to be maintenance free and made from materials that were stable. Here was an opportunity to use stainless steels, nickels, chromium, bronzes, glass and enamels. The Chrysler building was refurbished in the early 1980s, but the entrance required little work.

galvanizing and is also more porous, providing good adhesion for subsequent painting. In a similar way aluminium (aluminizing) and chrome (chromizing) can also be applied. See BS 4921:1988 *Specification for sheradized coatings on iron and steel articles.*

Anodizing   This is a treatment for aluminium. Although aluminium forms a continuous and strong oxide film which prevents further corrosion, the thickness of the film can be artificially increased by anodizing. Objects to be treated become *anodes* in an electrolytic cell, and the applied voltage causes aluminium ions to migrate outwards, towards the edge of the existing oxidized layer. Once they reach the edge they react with oxygen (and spare electrons from the dissociation of water) to produce more aluminium oxide (alumina). This thicker oxidized coating is quite porous and can be dyed to give colour prior to sealing. See BS 1615:1993 *Method for specifying anodic oxidation coatings on aluminium and its alloys* and BS 3987:1991 *Specification for anodic oxide coatings on wrought aluminium for external architectural applications.*

Controlled patinas   A great variety of treatments can be carried out with copper and its alloys as well as steel. They all involve treating the surface with various acid mixtures or particular compounds and then, after the surface has stabilized (often after wrapping for a period of at least 24 hours), waxing or sealing the surface. The nature of the finish is bound to be experimental and there will be considerable colour variations, although these variations add to the attraction. Experimental samples should always be made first for approval. Depending on the compounds used, pinks, greens, browns and yellows can be obtained with copper-based alloys. The effects on iron and steel are more limited.

Weathering steels Weathering steels are alloyed steels which contain a range of elements that will provide a coating to steel so that it will corrode no further. 'Cor-ten' is the best example of this: it is the American name for an alloyed combination of iron, carbon, copper and phosphorous which provides a particularly good oxidized coating. This type of alloy was introduced into the UK in 1967 and is known as '50 Grade weathering steel'. After a few years the steel weathers to a purplish-brown. But there are problems: although the material is widely used in the United States for engineering structures, as it saves on normal maintenance, different parts of the structure can weather differently according to how they are exposed, and how water runs down different sections. Staining can also occur and be unsightly and should be catered for.

Examples of metals used for exterior finishes in various ways are shown in Figs 6.4–8.

Figure 6.7 Lloyds of London (Richard Rogers, 1986). Here, the choice of metals is critical. Stable alloys have to be chosen which will not react with each other. Lack of surface coatings places heavier demands on engineering the right alloys and their connections.

Figure 6.8 Paolozzi sculpture. This small piece is leaning against a wall and belongs to Colin St John Wilson (Cambridge). The piece is rusting, but is robust enough to give years of pleasure before final decay (nature of metal unknown).

7 Composites

7.1 Introduction: Common applications of composites in finishes

We often use materials formed from several different materials bonded together. In any current book on materials science these new materials will be called *composites*. The behaviour of polymeric, ceramic or metallic materials in a composite changes completely. The new materials produced often have very different properties, including strength and durability.

When one material is combined with another giving a new range of properties, the whole new material can be regarded as a composite. Some materials already discussed in this book have their properties altered to such an extent they can be considered to be composites. For instance, concrete, which combines cement and aggregates, could be classified as a *particle* composite. When steel reinforcement is added, it then acts as a much stronger *fibre* composite. Renders and plasters may both contain strengthening glass fibres which change the properties of the original materials and, as fibre composites, they have greater flexibility and are more resistant to cracking. Polymers can yield a wide range of materials but, whether produced as thin surface coatings or more substantial sheet flooring systems, the basic polymer has a number of components added to modify properties and economise on the use of expensive resins (e.g. by using fillers). These finishes could also be regarded as particle or fibre composites on a smaller scale. Most finishes are bonded to their backgrounds in such a way that they act as one entity. This is why there is such a great emphasis on the preparation of substrates for finishes, since their effectiveness usually relies on good bonding.

There are many examples of composite structures in nature: leaf forms are composed of layers that give them great strength and resilience. Timber is a natural composite with a structure of cellulose fibres bound together with lignin that makes up the cell wall. As a fibre composite it has great strength and flexibility which allows it to resist quite extreme wind loadings while growing. In engineering it is a major structural material, ideal for boats and aircraft due to its strength and lightness. As a finish it has a major role in the building industry for external or internal cladding and for flooring materials. As a material, timber is already covered in MBS: *Materials* and *Materials Technology* although in this book the chief characteristics of decay are given in section 3.6.

Many other composites used in the building industry are manufactured. Some composites are made in order to extend the bulk of a high-cost material with cheaper additives or in order to modify the properties of a material for specialist use. The second purpose allows the manufacture of a great range of products that have a generic base or give an opportunity for recycling materials. For instance car tyres are recycled for rubber crumb which is used in flooring systems. Recycled PET plastic waste (polyethylene terephtalate) is now being assessed and may provide a liquid resin base for polymer concrete systems. It should completely replace cement and water as a binder. If this project is successful, it would have the effect of using up volume plastic waste and reducing the energy-intensive production of cement.[1]

7.2 Types of composite

A composite material is one made from two or more materials which give a range of properties and behaviour not found in the individual component materials.

There are three types of composites:

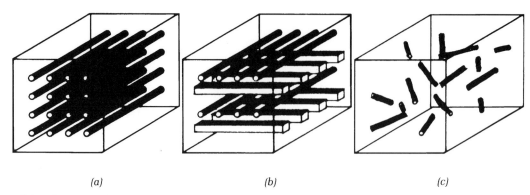

(a)   *(b)*   *(c)*

Figure 7.1   Fibre composites and directional reinforcement: (a) and (b) aligned; (c) non-aligned.

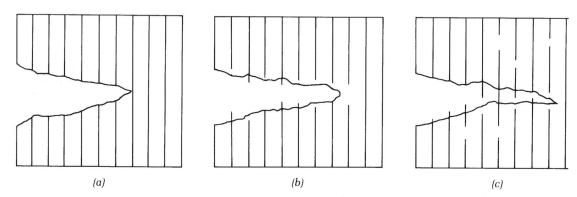

(a)   *(b)*   *(c)*

Figure 7.2   Fibre composites and crack propagation. Fibres can be brittle and set in a brittle matrix (a) or ductile matrix (b). The form of failure will be influenced by the composition of the matrix. Apart from straightforward crack propagation, there may be debonding and fibres may be pulled out (c). The form of failure will also be influenced by the nature of the interface between fibre and matrix. It is also possible to combine ductile fibres within brittle or ductile matrices.

(1) *Fibre*: Fibre composites are directional composites and can be designed to give directional strength to a material. Common examples are GRP (glass reinforced polyester) or GRC (glass reinforced cement) or even RC (reinforced concrete where steel acts as the fibre reinforcement). These composites are said to be 'anisotropic', i.e. they display different properties in different directions (Figs 7.1 and 7.2).

(2) *Particle*: Particle composites are 'isotropic' and display the same properties and strengths in all directions. Examples include: concrete with different particle sizes set in a matrix of cement, metals which are often alloyed with other metals or phases to produce a mixed anisotropic crystalline structure, cements which are a sintered mixture of ceramic and metallic phases used to produce cutting tools, brake pads, and HDPE (high density polyethylene with elastomer particles set in the parent matrix to increase

toughness) (Fig. 7.4). These composites are designed to maximize crack-stopping capabilities (Fig. 7.3).

(3) *Laminate*: Laminate composites are made out of thin strips or sheets that are bonded together. They often have enhanced properties in bending in one or more directions. Examples include plywood where veneers of timber are bonded together with epoxy resins to produce high-grade structural materials or laminated plastics where resins and fillers are bonded together using pressure and heat to produce new materials.

Rigid particle composites in flooring

The use of terrazzo finishes and other flooring systems which bed aggregates in a cementitious matrix is well tried and tested. There is a growing market for ready-made hard-particle composites of this nature in a large tile format. They are made by casting blocks of

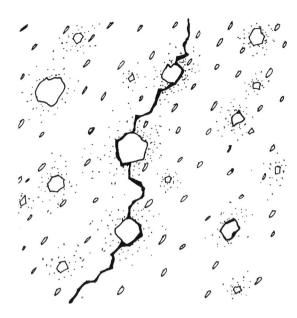

Figure 7.3   Particle composite. A common configuration for particle composites, whether in small-scale flooring systems, concrete or ceramics. Crack-stopping occurs because it takes energy to divert a crack around a particle.

aggregate material in a cementitious matrix, then sawing to the right dimensions with computer-operated diamond cutters that can produce sheets as thin as 7 mm.

There have been a number of failures with these flooring systems: the first signs are cracking and/or pull out of aggregate. These large scale tiles are often unstable for several reasons. The process of making the blocks and then sawing them induces stress. After the slabs are cut they are quickly delivered to the site; it can be a matter of a few weeks from manufacture to

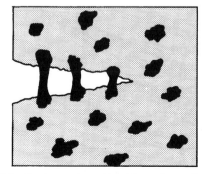

Figure 7.4   Crack-stopping with rubbery phases. Rubber particles are used to toughen certain plastics (high density polyethylene for example).

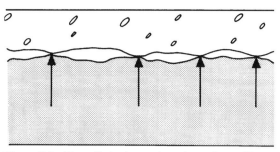

Figure 7.5   Floor tile inadvertently acting as beam. If a floor tile is not adequately bedded or there are protruding defects on the tile, it will have to span a distance and stresses will be incurred as point loads. Within artificially made conglomerates (particulate composites), there is no directional reinforcement, and cracking is likely on poor backgrounds.

site. We have seen that rock should season on exposure to the atmosphere before use, the aggregates in these materials have not been exposed to moisture etc. If the bedding of the flooring is uneven, this will place stress concentrators on the material and fractures will be induced (Fig. 7.5). If the slab matrix is resin-based, it will allow the tiles to bend a little under stress, but thinly sectioned aggregates will fracture in a brittle manner (Fig. 7.5).

7.3 Flooring applications

Floor coverings use ceramic, polymeric and composite materials. The whole floor assembly could be regarded as a composite; a combination of two or more materials and a substrate such that a larger-scale layered composite is formed. The whole construction will then work as one material. Flooring materials adhere properly to their substrate according to the principles of adhesion technology. Generally finishes that are used as floorings are to be found from each category of material in this book. For example, stone, tile and concrete floorings are listed in the ceramics section, and asphalt and vinyl floorings are listed in the polymeric section. However there are a number of other flooring materials, including wood, that are more easily identifiable as natural or manufactured composites and they are discussed in this section.

Floor coverings should successfully transfer load to the substrate without distortion or failure. The Japanese term 'shiki' refers to the notion of force and the 'tatami' mat (often 50 mm in thickness) is seen as a way of successfully spreading force to the substrate. Concentration of force should be avoided and the concentration of mass on point loads can create enormous pressures. For instance stiletto heels used to produce pitted surfaces in timber and cause permanent damage. Floorings that have some 'give' and transfer load gradually are more comfortable to walk on and transfer less shock through the spinal column than hard surfaces. Even though most of our floor systems are rigid, this transference of load from one layer to another is even more important to control as, ironically, the strength of rigid systems is illusory and they are more prone to brittle and sudden failure.

The British Board of Agrément in its MOAT (Method of assessment and testing) Series 2 gives directives for the assessment of floorings. This is an essential reference text listing all types of flooring and covering the complete range of parameters which assist in choosing the right type of covering material. In MOAT series 23, directives are given for the specification of plastic floorings.

The most important component of a flooring is the base and Table 3 of BS 5385 summarizes the suitability of flooring beds for different bases (Table 7.1). Generally only screeds (more than 3 weeks old) and concrete bases (more than 6 weeks old) that have cured are suitable to take mortar beddings or adhesives for floor finishes. Timber and asphalt are normally unsuitable as deflection and movement may be too great to prevent flooring failure. To ensure that timber floors are adequate to take ceramic type finishes they should have noggins fixed every 300 mm and be surfaced with exterior grade plywood at least 15 mm thick which is fixed every 300 mm. These heavier grade floor finishes should then be fixed with adhesive. The final thicknesses and suitability of flooring beds are listed in Table 7.2 and the recommended thicknesses and mix designs of floors are listed in Table 7.3.

Most buildings have floor finishes applied on screeds or slabs which are fully described in BS 8203:1987, which gives a detailed guide to their construction.

There are four ways of laying slabs to receive floors:

- Finished slab with surface adequate to receive flooring.
- Slab with screed applied while slab is still green, i.e. within 3 hours of placing. Otherwise known as *bonded slab* as the setting of screed and slab can take place simultaneously. The screed should have a minimum thickness of 25 mm.
- Slab with screed applied after the slab has initially set. This is an *unbonded slab*. The screed is sufficiently thick to maintain individual structural integrity.

Table 7.1 Suitability of flooring beds for different bases

Base	Cement:sand mortar and cement:lime: sand mortar bonded to the base	Cement:sand semi-dry mix		Adhesives
		Bonded	Unbonded*	
New concrete (less than 6 weeks old)	U	U	S	U
New screed (less than 3 weeks old)	U	U	S	U
Mature concrete	S	S	S	S
Mature screed	S	S	S	S
Screed over suspended floor or underfloor heating	U	U	S	C
Suspended in situ concrete				
Rigid and new (less than 6 weeks old)	U	U	S	U
Rigid and mature	S	S	S	S
With significant deflection	U	U	S	C
Timber	U	U	U	C
Asphalt	U	U	S	C
Existing hard floor finishes after preparation				
Terrazzo	S	U	S	C
Unglazed ceramic tile	S	S	S	S
Glazed ceramic tile	S	U	S	C
Granolithic topping	S	S	S	S
Natural stone	S	S	S	S

Note: S – 'Suitable'; U – 'Unsuitable'; C – 'Confirm' suitability with the manufacturer or supplier.
* Unbonded beds are generally unsuitable for heavy traffic conditions.

Source: Based on Table 3 BS 5385 Part 5:1994

Table 7.2 Flooring beds: final thicknesses and suitability

Type of bed	Thickness (mm)		Terrazzo tiles and slabs	Limestones and sandstones	Marbles, granites, slates and other stones	Composition block
	Minimum	Maximum				
Cement:sand mortar						
Generally	15	25	S	C	S	S
Flooring units less than 10 mm thick	10	15	N/A	U	C	N/A
Flooring units of variable thickness	20	30	N/A	C	S	N/A
Cement: sand semi-dry mix	25	70	S	S	S	U
Cement:sand semi-dry mix over a separating layer	40	70	S	S	S	U
Cement:lime:sand mortar	15	25*	U	S	U	U
Adhesive	1	6†	C	C	C	S

Note: S = 'Suitable', U = 'Unsuitable', C = 'Confirm' suitability with tile manufacturer or supplier.
* May be applied up to 50 mm thick for large flooring units.
† Some may be applied up to 12 mm thick for filling small isolated depressions.

Source: Based on Table 4 BS 5385 Part 5:1994

Table 7.3 Recommended thicknesses and mix designs

Case	Thickness (mm)	Total cement: aggregate	Mix proportions by weight of dry material				
			Cement	Sand	3 mm aggregate	6 mm aggregate	10 mm aggregate
Light duty	8–15	1:3 to 1:4.5	1 1	3–4.5* to 3–4†	–	–	–
Medium duty	10–15	1:3–4.5	1	1.5–2.25	1.5–2.25	–	–
	15–30	1:3.5–4.5	1	2.0–2.25	–	1.5–2.25	–
	25–40	1:4–5	1	2.25–2.50	–	–	1.75–2.50
Heavy duty	As for medium duty	As for medium duty except that very hard single-sized aggregates should be used, as given in 5.3.					

* Grade C in Table 4 of BS 882:1992.
† Grade M in Table 4 of BS 882:1992.

Source: Based on Table 1 BS 8204 Part 3:1993

- Slab with a floating screed laid over insulation which has to be of sufficient depth, usually at least 75 mm, and possibly reinforced to take light loadings.

The surface of a floor, whether screed or slab, should have tolerances specified so that the floor surface finish can be laid properly. BS 8203 states that a high standard would not have a surface variation greater than 3 mm. In most cases a tolerance of 5 mm would be regarded as normal and 10 mm would only be satisfactory for heavy duty, 'utility' areas.

Floors are classified by the type of use expected and assigned a letter as shown in Table 7.4.

Fire properties Although there are no formal requirements for fire resistance or incombustibility, floor surfaces should, where possible, have the same fire classification as the building. Floor surfaces for

Table 7.4 Flooring grades

Code	Flooring grade
HI	Heavy duty industrial: can take heavy vehicles, impact loads in factories, dense loads
LI	Light duty industrial: lightweight trucking, carriage of low density materials
HP	Heavy pedestrian: corridors in schools, colleges, barracks
NP	Normal pedestrian: residential and domestic buildings, small classrooms and offices
RS	Roller skating
G	Gymnasia
CA	High permeability to chemicals and acid
FH	Suitable for floor heating
B	Ballrooms and dance floors
D	Decorative
S	Small movement

escape routes should be non-combustible. At a very minimum they should have the same fire rating classification as the building. Notice should also be taken of smoke and toxic gases that can be emitted from the burning of polymeric materials (see BS 8203). Standards usually represent a minimum requirement. But this is one area of risk where economies should not be made. Buildings exempt from formal regulatory approval should still adhere to these standards.

Further reading: Readers should consult *Fire Test Results on Building Products: Fire Propagation* by R.W. Fisher, and by the same author, *Fire Test Results on Building products: Surface Spread of Flame*. Both are published by the Loss Prevention Council.

International Wool Secretariat: *Flame Resistant Treatments*.

B.S. 3119 *Methods of test for flameproof materials*.

Freedom from slipperiness   Falls and injuries due to slipping cause serious concern in the UK. In 1979 5,895 fatalities and 0.5 million injuries were due to falling. Slipping accounted for 50% of the injuries. Such falls account for a great deal of absenteeism and designers must minimize the risk.

In order to measure how slippery a surface may be, one measures the frictional resistance of a surface (mu). There are few guidelines. On the Continent one test uses a subject standing with wet feet on a surface which is gradually tilted until slip occurs. From this test a friction coefficient can be calculated. Remember that these coefficients relate to a pairing of surfaces, and there are great differences between wet feet, rubber-soled and leather-soled shoes. Low values of a coefficient indicate high slippery areas.

Values for ordinary floors range from approximately 0.1 to 1.0. PTFE (polytetrafluoroethylene) has a very low value of about 0–0.04 with many surfaces. Wood on most surfaces has values of approximately 0.4, steel on steel has a value of about 0.6. Rubber on tarmac has a high value of about 1.0, but rubber on ice has a value of 0.1, hence the importance in keeping entrance ways dry and warm. Greasy surfaces can bring the value down dangerously to about 0.1 so correct cleaning and maintenance become an important part of flooring technology.

The GLC in their Development and Materials Bulletin No 145 (Item B2 *Sheet safety surfaces for floors: Methods for evaluation*) itemize activities and discuss the mechanics of frictional resistance and 'slipping'. They used a TRRL skid resistance tester which had a scale of calibration from 0–150 indicating the degree of resistance to slipping. Below 20 was considered dangerous, between 20 and 39 marginal and therefore unacceptable, between 40 and 74 satisfactory, and above 75 excellent and essential for heavy traffic in public areas. The most difficult case of

Table 7.5 Categories of hardwood and softwood timber

Timber type	Density (kg/m^2)	Category
Hardwood		
Brush Boxwood	930	HI, RS, CA
Maple rock	735	HI, HP, B, G, RS
Oak (European)	705	D, HP, CA
Oak (Tasmanian)	610–710	G
Hornbeam	735–770	HP
Walnut	670	D, NP
Yellow Birch	690–710	NP, LI, B
Softwood		
Douglas fir	495–530	LP, G
Pitch pine	655	NP
Redwood, Scots pine	480	LP
Whitewood	465	LP, CA
Pine (American)	530	LP

Table 7.6 Timber flooring applications

Type of composite or processed timber	Application	Sizes and indicative price per square metre (1994)
Boards* Softwood planks normally butt-jointed, can be tongue and grooved; can be left natural and sealed, or painted; more likely to receive carpet Cork or vinyl tiled finishes should be laid on an additional hardboard layer BS 8201:1987 BS 1297:1987 *Grading and sizing of softwood flooring*	Boards should be well cramped and laid over joists at centres to provide safe loading conditions. See current Building Regulations for Tables Netting can be laid over joists to hold indulation on suspended ground floors Moisture content to be 12–15%	Widths: 65–137 mm; thicknesses: 16–28 mm Price: £15–24 depending on size and finish
Strips* Hardwood and softwood often provided in packs, pre-sealed or polished with tongued and grooved edges and ends BS 8201:1987	Often supplied as random bundles with fixing systems that allow for secret nailing over a separating foam layer; 19 and 25 mm thicknesses can be fixed direct to joists.	Widths: 50–102 mm; thicknesses: 6, 9, 12, 19 and 25 mm Price: £24–30 for domestic work; £35–50 for commercial, depending on timber species
Blocks* Hardwood and softwood, tongued and grooved BS 8201:1987 BS 1187:1959 *Wood blocks*	Blocks can be laid in hot bitumen or with adhesive directly onto screeds Pitch pine blocks are laid with end grain exposed upwards.	Sizes: 229 × 75 × 21 to 305 × 75 × 21 mm Softwood (pitch pine) resemble cobbles; standard lengths: 102, 127 or 229 mm all with a width of 76 mm, but thicknesses of either 63 or 114 mm Price: £16–45 depending on species (Iroko and selected oaks)
Parquet* Hardwood strips either butt-jointed or tongued and grooved; panels are usually parquet mounted on laminated softwood. BS 4050 Part 1:1977 *Specification for mosaic parquet panels* & Part 2:1966	Individual parquet can be fixed over existing softwood boards but these should be overlaid with 4 mm plywood or 4.8 mm hardboard Panels can be fixed with adhesive straight onto screeds or bedded in bitumen, both methods can use secret nailing through tongues	Traditionally brick shaped but lengths vary; widths: 51, 76 or 89 mm, generally 5 mm thick increasing to 10 mm for larger sixes Can be laid in pre-made panels 305 or 610 mm square, 25 or 32 mm thick Price: £20–24 for residential, 40–70 for commercial
Mosaic* Thin strips, less than 15 mm width laid in weave pattern often with backing and protective plastic facing stripped off after laying BS 8201:1987 BS 4050:1968 *Mosaic*	Can be fixed onto screed as for blocks or laid over overlay on softwood boards	Produced in panels 305 or 457 mm square, and 10 or 16 mm thick respectively

continued

safety they considered was standing on tube or railway platforms.

Resistance to sparking Sparks are caused by static electricity and not flowing charge from a supply. Anti-static flooring is required in office buildings where equipment such as magnetic relays, circuitry and data storage can be affected.

Raised floors Raised floors are becoming standard in offices where deep servicing is needed for information technology and communications. Large format tiles are laid either on jacks or on continuous plastic or timber runners. The jack system may allow for easier levelling especially in existing premises and is possible on unscreeded floors, although installation may prove more expensive due to the additional time

Table 7.6 continued

Type of composite or processed timber	Application	Sizes and indicative price per square metre (1994)
Plywood† BS 6566 Part 1:1985 *Specification for construction of panels and characteristics of plies, including making* BS 6566 Part 6:1985 *Limits of defects, classified by appearance* BS 5669 Part 2:1989 *Structural plywood* BRE IP 4/85 (MR: Moisture resistant, CBR: Cyclic boil resistant, WBP: Weather and boil proof)	Face grain laid at right angles to structural supports Fix with 50 mm nails at 150 mm centres, unless used as structural deck with joists, when it should then be bonded and screwed to engineer's specification Fix 13 mm thickness at 460 mm centres, 16 mm at 610 mm centres. Bond class MR unless risk of damp when CBR should be used or WPB for heavier duty with risk of wetting.	2440 × 1220 × 4, 13, 16 or 25 mm in thickness; the 4 mm thickness can only be used as an overlay, fully supported. Price: £18–20
Chipboard† BS 5669 Parts 2 & 5:1989 *Resin bonded wood chipboard* BRE IP 3/85 Can be tongue and grooved C1–C5 grades-structural strength	50 mm nails at 400 mm centres C4, C5 should satisfy most purposes Fix 18 mm thickness at 40 mm centres, 22 mm at 600 mm centres	2440 × 1220 × 12, 18 or 22 mm in thickness; the 12 mm thickness can only be used as an overlay, fully supported Price: £8–12
Hardboard† BS 1142 Part 2:1989 *Fibre building boards* BRE IP 12/91 (S: standard grades; T: tempered)	Used as overlays to timber to receive flooring finish. Use panel pins at 200 mm centres on lines 400 mm apart Bond to concrete with adhesive Grades SBN, SHA, SHB, SHC; use SBI, THE, THN where risk of wet conditions	2440 × 1220 × 4.8 or 6 mm; 4.8 mm thickness can only be used as an overlay, fully supported Price: £4–7
Cork Natural cork particles and resins, sheet material on mesh; natural buff colour, also stained to various browns, greens and reds; available as sheet or tiles Resistant to oil, occasional water and weak acids, but not alkalis BS 6826:1987 BS 8203:1987	Use proprietary adhesives, lay on overlay hardboard or direct to plywood or chipboard Keep at room temperature for two days before laying and then apply rolling pressure after bonding	Width of 1830 mm and lengths up to 15 000 mm; 3.2–6 mm thickness for rolls Price: £11–22 for residential, £15–30 for commercial
Linoleum Powdered cork, mineral fillers, linseed oil and resins with pigments giving a wide range of colours, compacted under high pressure on to woven backing or cork; resistant to oil and weak acids, not alkalies or continual damp BS 6826:1987	Use proprietary adhesives, lay on overlay hardboard or direct to plywood or chipboard Keep at room temperature for two days before laying and then apply rolling pressure after bonding Patterned linoleum is laid loose often over paper felt	Widths: 1830 or 2000 mm; thicknesses: 2–6 mm for sheet Square tiles: 229, 305 and up to 915 mm; can be 3, 5, 8 or 25 mm thick Price: £15–30

continued

taken. There should be firebreaks in these flooring systems especially where partitions above form compartments. As part of the fire protection system it may be seen only to flood these areas with halon gas in the event of fire. One unexpected repercussion of deep floor systems is the provision of new warm homes for the city of London's fluctuating rat population. Infestation should be attacked.

The extensive local cabling needed for networking is a problem where space is not available for deep floor

Table 7.6 continued

Type of composite or processed timber	Application	Sizes and indicative price per square metre (1994)
Composition blocks Made from mixture of mineral fillers (chalk), wood granules, pigments with cement, PVC or oil binders and water-pressed into blocks and cured Limited colours, good resistance to chemicals and oils BS 5385 Part 1:1990 *Wall and floor tiling* Part 2:1991 (Code of Practice); this standard gives good construction details for floors and skirtings.	Installed on screed or concrete with 13 mm bed of 1:3 mix (cement:sand) or thin-bed adhesive	$157 \times 52 \times 10$ or $190 \times 63 \times 16$ mm Price: £50–55 for 174×57 mm blocks

Note: Timber flooring can be stained, sealed and polished. Strip flooring can often be supplied ready sealed. Maintenance requires renewal of polished finishes with the advantage of sanding to expose a new surface after years of wear. Blocks are extremely hardwearing due to their density.

* BRE Digest 323, January 1992 Wood-based panel products contains good diagrams, and information on grades of boards for different applications. Grades must be specified in practice relating to bonding, durability and structural strength. The bonding type is critical if there is a risk of exposure to damp.
† All of these finishes should be cleaned with neutral detergent solutions and can take a light polish. They are durable finishes, have some resilience and are warm finishes. Blocks can be sanded after long use to give a fresh surface. As well as having greater resilience, cork acts as a good barrier to sound transference if over 5 mm in thickness and is a non slip finish.

systems. A German manufacturer, Herforder Teppiche, has developed a composite flooring system only 16 mm deep made of concrete and epoxy resin. A studded profile, similar to small-scale egg boxes allows for cables to be threaded and the studs support steel sheets which then take the normal flooring finish.

Timber flooring Table 7.5 lists temperate hardwoods and softwoods from sustainable sources available in the UK. These are also listed in FPRL Bulletin 40 with letters denoting the category of use. The British Standard lists tropical hardwoods as well, but specifiers should ensure such woods are labelled with the country of origin and conform to EC regulations on supply from well-managed and sustainable forests.

Table 7.6 lists natural timber flooring applications.

Note

1 *Mechanical Properties of Polymer Concrete Systems Made with Recycled Plastic* Rebeiz, Karim S., Fowler, David W. and Paul, Donald R. *ACI Materials Journal* Jan–Feb 1994. Optimum designs used 10% resin, 45% oven-dried coarse aggregate, 32% over-dried sand and 13% fly ash.

Conclusion

Although this book deals with a range of available materials and observes that their choice is generally to do with ultimate performance and minimization of degradation, there are some other factors worth considering in making choices of materials.

Healthy occupancy It is now acknowledged that indoor pollution in buildings can make people unwell by creating a slightly toxic environment. Although articles had been published in the scientific press outlining some major pollutants, e.g. background radiation, nitrous oxides and gases from formaldehydes (*New Scientist* 5.12.85 *The Aggressive Environment*) the issue did not reach the newspapers until 1987. Poor ventilation is thought to be a major contributor. This is partly a result of energy-saving in well-sealed buildings with a highly controlled environment. However, highly controlled environments that minimize air movement and convective forces can endanger health.

Some new materials have also been found to be dangerous. In 1984 Wanner[1] reported, at the Third

Figure 8.1 Summit plateau, Ben Nevis. (*Photograph*: Steve Keates.)

International Conference on Indoor Air Quality and Climate in Stockholm, that emission rates of formaldehyde from particle boards in some new buildings, after only one year, have been found to exceed admissible limits. Uncoated panels could continually emit formaldehyde which would initially give symptoms of a sore throat, irritation to the eyes and feelings of general discomfort. People are able to detect concentrations of formaldehyde below 1ppm and there have been more cases of dermatitis and asthma in ultra-formaldehyde foam insulated houses in the USA. Long term exposure can cause severe respiratory disease.

Formaldehyde vapour has for some time been confirmed as an irritant but, in the building industry, observations have been confined to the use of urea formaldehyde foam insulation in cavity walls. In BRE Information Paper No 25 (1982) observations are made that although it has been in use for over 20 years it is now known that it can cause discomfort through 'irritation to the eyes and respiratory tract'. It is categorically not advised for lightweight timber construction and other systems that have a high level of through ventilation, but only in sealed systems, e.g. in brickwork cavity walls which consciously minimize the possible leakage of fumes. See BRE Digest No 236 for advice and also BS 5617:1978 which is the relevant British Standard. However, even in brickwork cavities, if the temperature of the walling system is raised sufficiently, the foam can deteriorate and release formaldeyde.

Indoor contaminants include

Asbestos (fire retardants)
Ozone (photocopiers, electrostatic air cleaners, electrical equipment)
Nitrogen dioxide (gas cookers)
Respirable particulates (dust)
Radon (soil, masonry, concrete, water service)
Formaldehyde (particle boards, plywoods, laminates, carpets)
Carbon monoxide (cigarette smoke, gas, coal)
Urea resins (laminates)
Ions
Allergens (e.g. mites)
Pesticides (timber treatment, eradication generally)
Microbes

Volatile organic compounds (paint)
Fibreglass (insulation)
Lead (paints)
Fluorocarbons, odorants (aerosol sprays)

Further reading: Turiel, I., *Indoor Air Quality and Human Health*, Stanford University Press, 1985.

Resources Specifiers should also allow for the sensitive use of resources. If materials are available in a form that do not represent a drain on resources for that material, they should be used. If the material is in short supply as it is a finite resource, or it is known that the use of it will in fact cause an unacceptable alteration of the environment, then it should not be used. There are some very significant areas of consumption which show a rising use of existing resources. For instance, the use of limestone is increasing, not just for cement and lime products, but for road stone and railway ballast, powdered for agricultural use (altering the pH of soils to alkaline conditions) and as a flux for the iron and steel works in Consett and Teeside. Although the useable bed of Limestone in Weardale is about 22 m thick the demand for very hard Carboniferous limestone is greater than for the softer and more porous Magnesian limestones, and the volume now being excavated is altering the landscape of Britain. The excavation of other bulk materials is also rising. Sand and gravel excavated has risen from 2×10^6 cubic metres in 1965 to 4×10^6 cubic metres in 1975. (British Regional Geology series).

If a material requires a lot of energy for processing with consideration of accompanying pollution from the products of combustion, then the material should not be used. There are situations where compromise may be unavoidable but in a world of diminishing resources our decisions should be qualitative and conscientious. After all, we are not just agents for others but powerful decision makers in the large scale use of materials, and the effects of bad decisions are often irreversible.

Note

1 Wanner, H.U., *Environment International* Vol 12(1–4) 1986, containing papers from Third International Conference on Indoor Air Quality and Climate, Stockholm, Sweden, 1984, pp. 311–315.

Index